Fundamentals of Pipe Drafting

CHARLES H. THOMPSON

Associate Professor

Head of the Drafting and Design Department

The Technical Institute

College of Engineering

Oklahoma State University

JOHN WILEY & SONS, INC.

New York · London · Sydney

Copyright © 1958 by John Wiley & Sons, Inc.

All Rights Reserved.

Reproduction or translation of any part of this work beyond that permitted by Sections 107 or 108 of the 1976 United States Copyright Act without the permission of the copyright owner is unlawful. Requests for permission or further information should be addressed to the Permissions Department, John Wiley & Sons, Inc.

Library of Congress Catalog Card Number: 58-13471

Printed in the United States of America

ISBN 0 471 85998 2

Fundamentals of Pipe Drafting

Contents

1 Introduction 1
2 Piping Symbols 3
3 Methods of Representing Piping Symbols 17
4 Diagram Drawing 20
5 Controls 23
6 Pipe and Pipe Fittings 33
7 Specification of Parts 41
8 General Arrangement and Diagram Drawings 47
9 Detail Drawing 51
Appendix. The Language of Piping 61
Index 65

Preface

This book is designed for students who have accomplished the basic skills of mechanical drawing and wish to develop specialized skills in the field of pipe drafting. Although many textbooks relating to pipe in its various uses have been published, very little consideration has been given to drafting in relation to pipe work. Many drafting textbooks touch on the subject, but for the student who desires to specialize more complete treatment is necessary.

It is my purpose to present information pertaining directly to piping as related to drafting. I have intentionally omitted subject material of special use to the engineer in order to concentrate upon the many details that the draftsman must take care of in his capacity as aide to the engineer.

The contents of this book represent what I have found, after ten years of teaching the subject of pipe drafting, to be the type of information from which the student draftsman of limited experience can best benefit. Although no book can substitute for a good teacher, I hope that many minor details so often overlooked in textbooks and made the responsibility of the teacher have been discussed in a way that will ease the load of teaching.

CHARLES H. THOMPSON

Stillwater, Oklahoma
August 1958

1 Introduction

Among the many fields of drafting, pipe drafting stands out as one in which the draftsman has great opportunities. A knowledge of the uses of pipe and the best methods of representation not only offers an additional field of opportunity but also serves to make the draftsman more useful to his employer.

The ability to do pipe drafting depends upon knowledge of several other kinds: machine drafting because of the large amount of machinery and equipment involved; architectural drafting because many piping problems involve architectural plans; and, in some cases, map drafting because pipe lines that cover large areas involve the use of maps. In addition, there are many special problems typical of pipe drafting that vary in difficulty from the selection of a simple pipe connection to calculations of fluid flow, pipe sizing, and stresses in pipe.

For some reason, probably because the opinions of what constitutes the requirements of a piping draftsman are varied, standard practices in pipe drafting have been slow in developing. Common practice in one company may be unacceptable in another; however, the rules of projection as well as all the common practices of machine drafting remain valid. No person can be successful if he is not thoroughly familiar with the common rules of mechanical drawing and lettering. A good general knowledge and skill in mechanical drawing coupled with the ability to use data supplied by the manufacturer are the basic skills required of the piping draftsman.

The piping draftsman should keep in mind that his first duty is to represent pipe installations and details as the designer wants them. The many engineering problems involved in pipe design are the responsibility of the engineer; however, once the designs are completed, it is the responsibility of the draftsman to follow the engineer's instructions and to put on paper in the form of drawings information the workman can understand.

The step between the completion of the design and the completed, detailed drawings is a long and important one, and one for which the draftsman is responsible; he must make the proper selection of standard parts, and arrange the design with respect to convenience, clearances, accessibility, etc. He is also responsible for the application of common practice in design established by handbooks, catalogs, and company literature.

2 Piping symbols

Symbols are the language of the piping draftsman. Each symbol taken separately is comparatively simple and easy to draw, but it must be remembered that every symbol has a definite meaning, as has every line. Careless work is apt to lead to costly errors.

SINGLE-LINE SYMBOLS

Single-line symbols are usually used on small-scale drawing, frequently as small as $1/8'' = 1'$-$0''$; however, different companies may vary the scale recommendations. Also, the purpose for which the drawing is to be used determines not only the scale, but also the completeness of detail. Drawings that are intended simply to show processes or principles of operation without application to definite structural situations are called diagram drawings and lend themselves well to the single-line method of representation. The single-line method of representing piping and symbols is also used when speed of execution is important.

Figures 2.1 to 2.4 illustrate the American Standards Association (ASA) recommendations for single-line symbols.

When drawing single-line symbols, usually no attempt is made to scale the size of the symbol to correspond with the size of the valve or fitting represented. There are no rules definitely regulating the drawing size of the symbols. On the average-size drawing, the symbols are usually drawn about 50 per cent larger than the symbols illustrated in Figs. 2.1 to 2.4.*

*Scaled piping templates are available from most drafting supplies sources. Their use greatly reduces time and effort in the drawing of piping symbols.

DOUBLE-LINE SYMBOLS

On drawings of large scale, or when accurate scale and detail are desired, a double-line method of representation is used. In this method, double lines are drawn and spaced to represent the scaled nominal outside diameter of the pipe.

Valve symbols used on double-line drawing closely resemble the symbols used in single-line drawing, except that flanges are drawn to scale, and the face-to-face dimensions on the flanged valves correspond to the actual scaled dimensions of the valve. Also, flanged and welded fittings are drawn to scale. Threaded fittings and valves are scaled only if there are close clearances.

Figure 2.5 illustrates some of the common symbols used in double-line pipe drawing. Note that threaded fittings are drawn to a close approximation of the actual shape of the fitting.

Frequently, when appearance is important, as in catalog illustrations, display drawings, sales drawings, or advertising literature, pipe, fittings, and valves are drawn to give as near a realistic appearance as possible. Examples of this type of drawing are shown in Figs. 2.6 to 2.10.

Due to the more intricate details of double-line drawing, the draftsman should not attempt to use a scale smaller than $1/4'' = 1'$-$0''$, and this only if the pipe sizes are large enough to show up well on the drawing. Excluding the types of drawings mentioned in the preceding paragraph, double-line drawings are usually made for the purpose of giving more detailed information than can be shown on single-line drawings of smaller scale. Many companies use a standard of

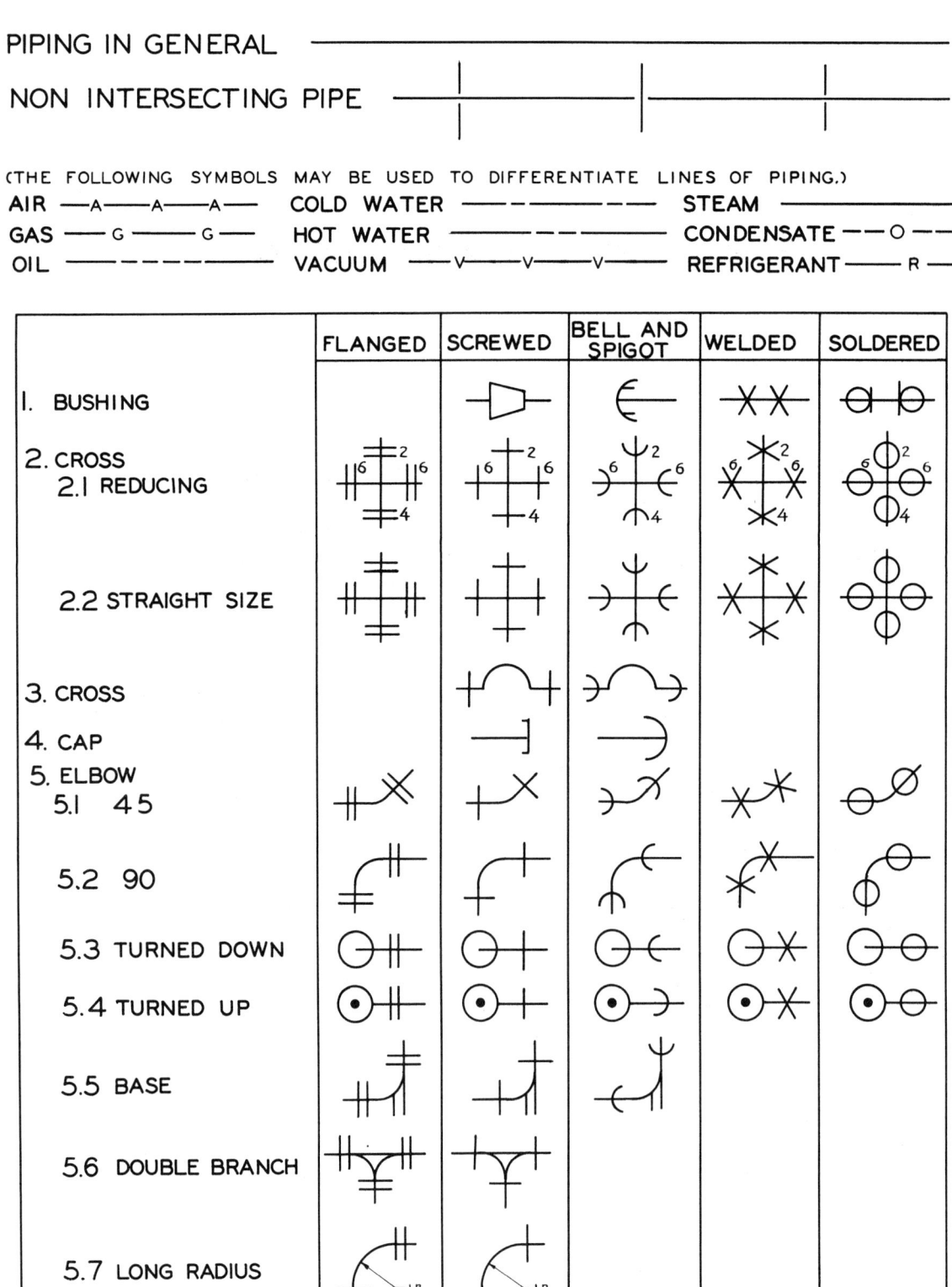

Fig. 2.1. Graphical piping symbols.

PIPING SYMBOLS

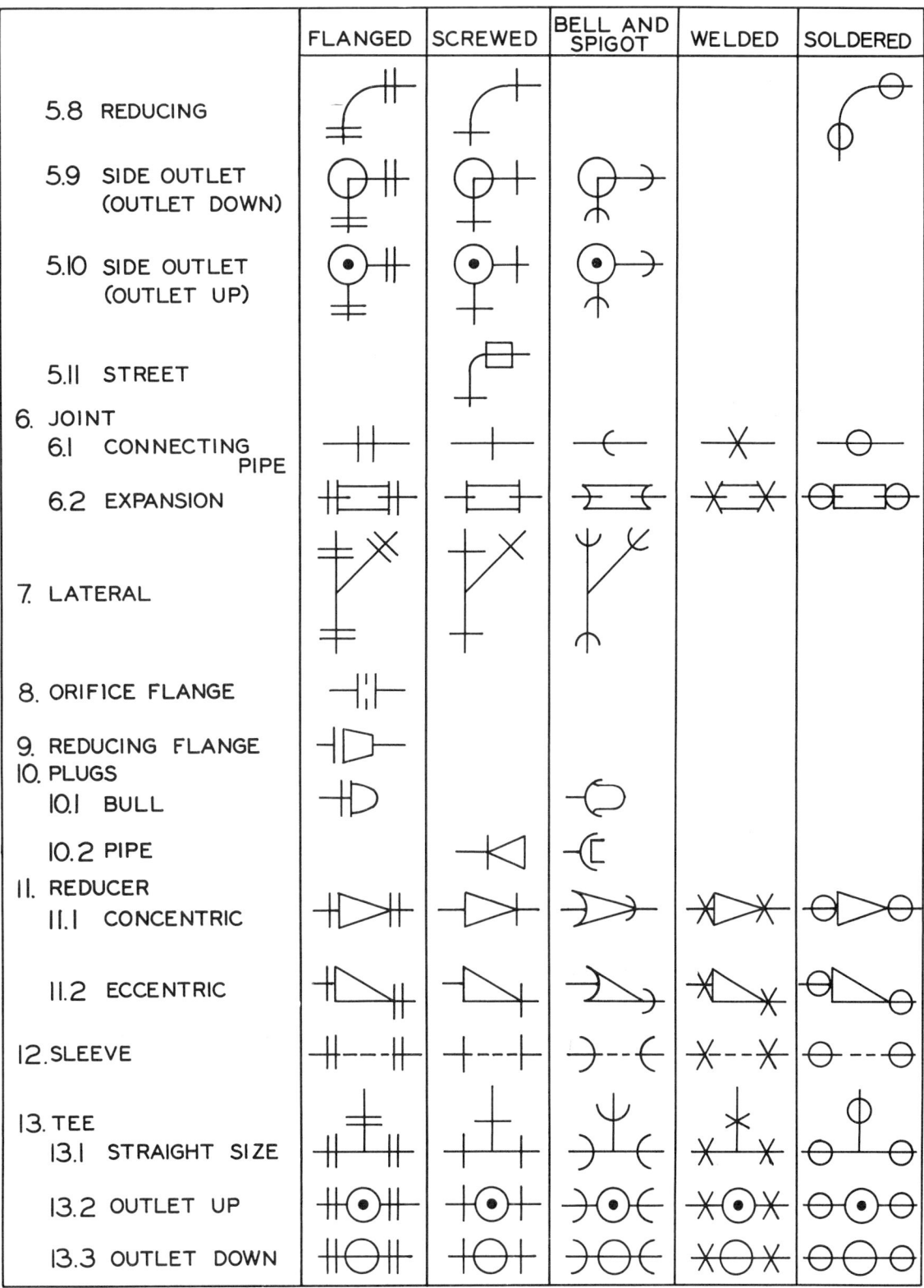

Fig. 2.2. Graphical piping symbols *(continued)*.

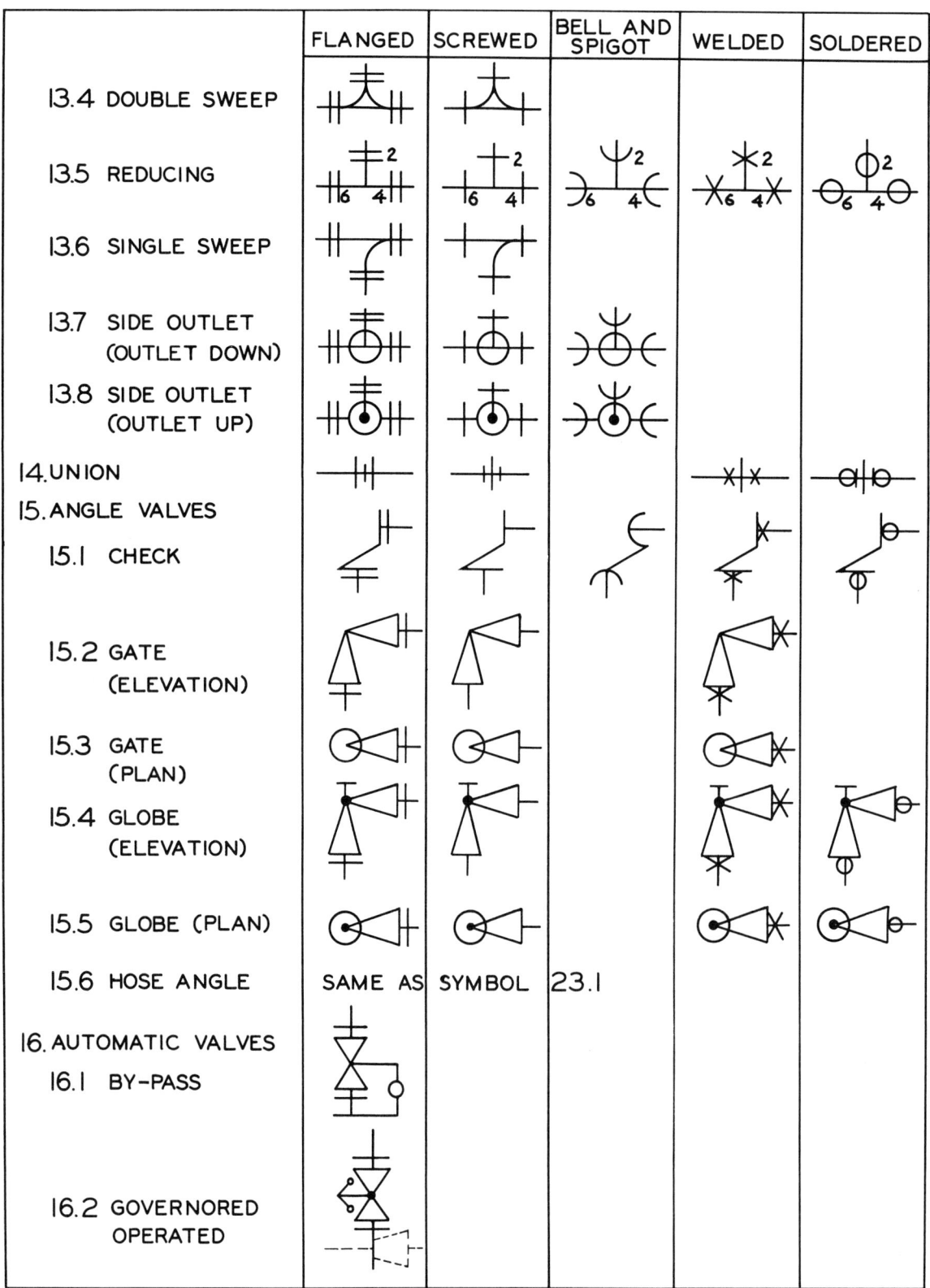

Fig. 2.3. Graphical piping symbols *(continued)*.

PIPING SYMBOLS

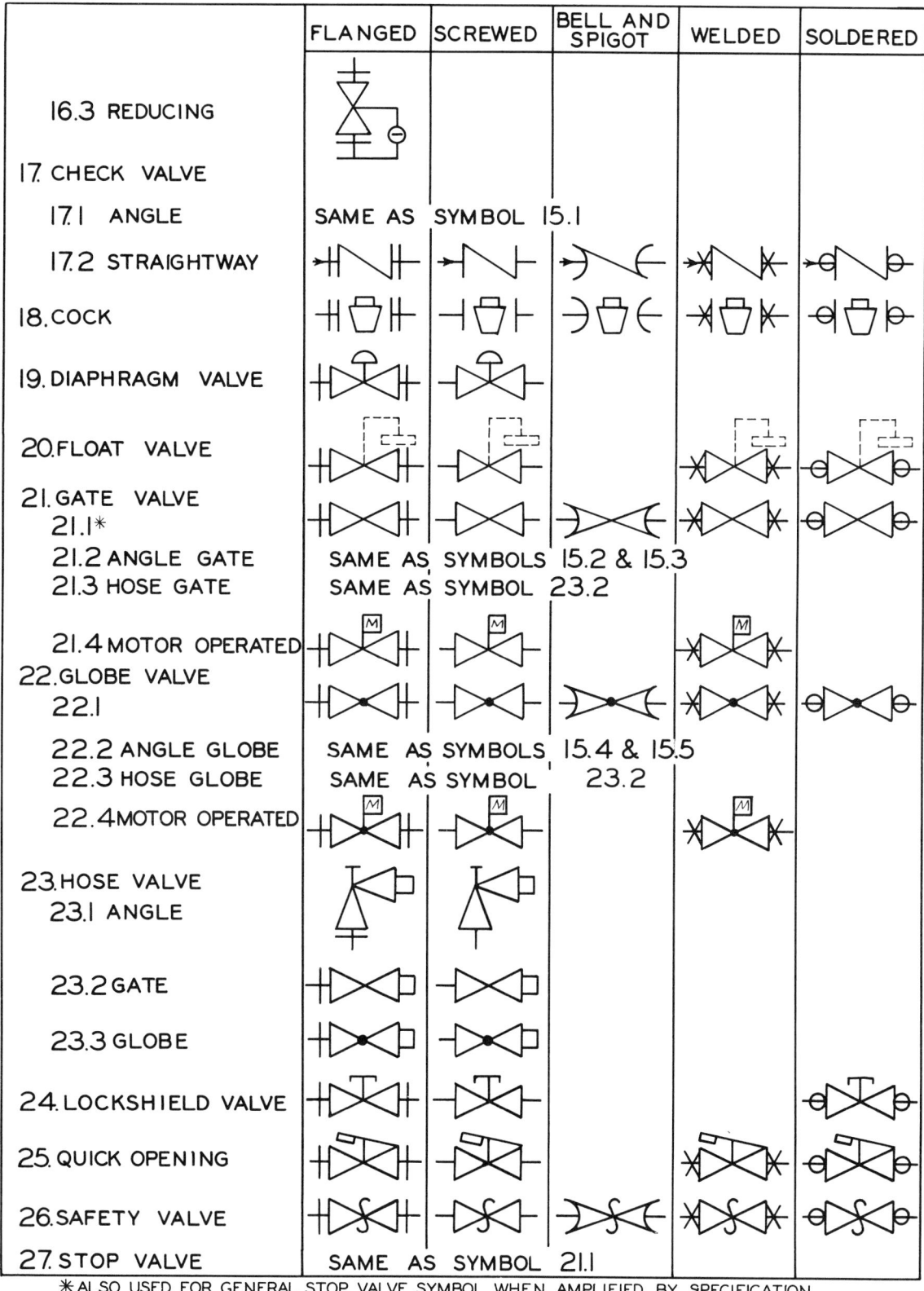

Fig. 2.4. Graphical piping symbols (*continued*). (Figs. 2.1–2.4 abstracted from *Graphical Symbols for Pipe Fittings, Valves, and Piping*, ASA Z 32.2.3–1949, published by American Society of Mechanical Engineers, New York.)

FUNDAMENTALS OF PIPE DRAFTING

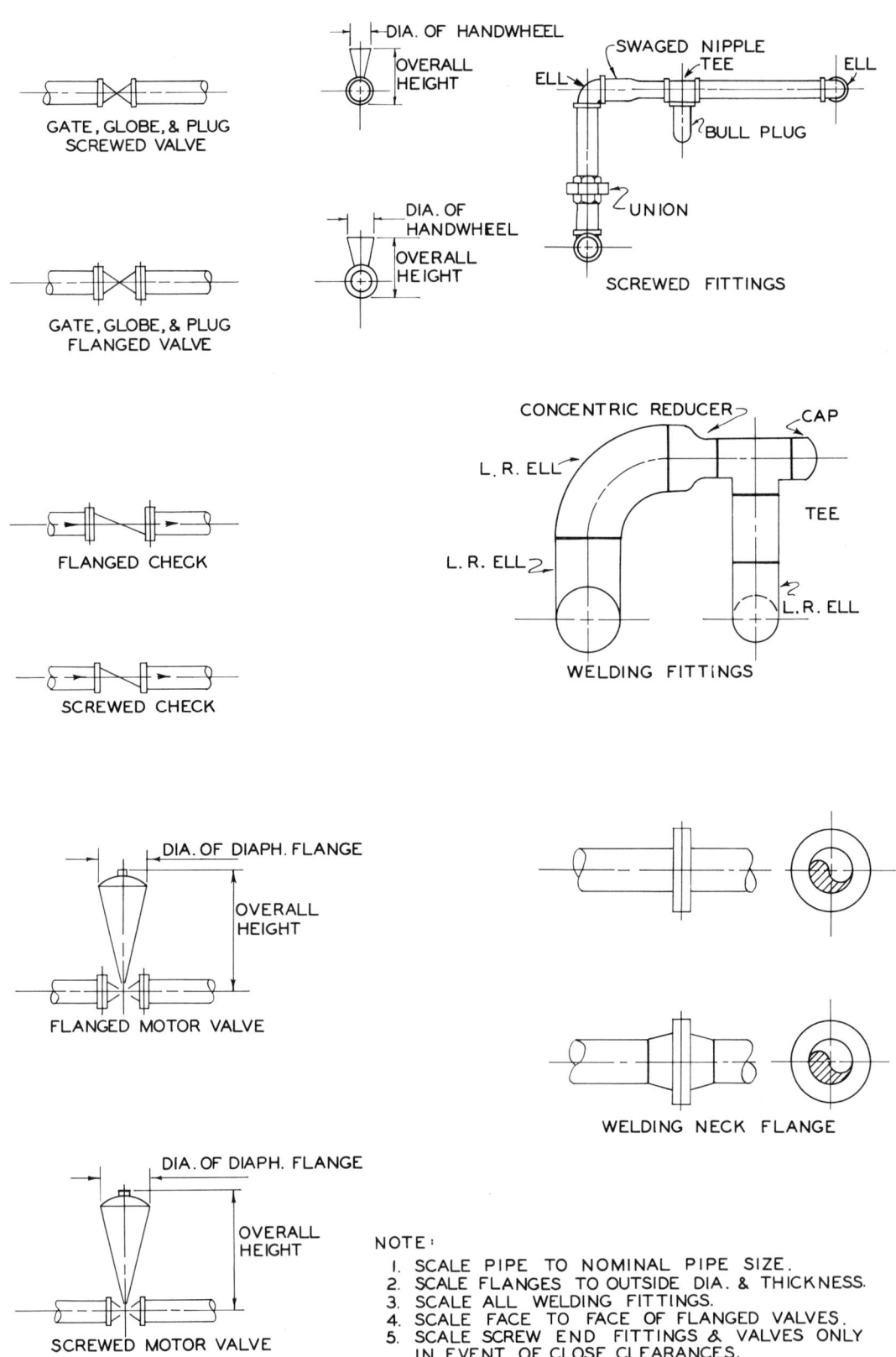

Fig. 2.5. Simplified double-line piping symbols. (Courtesy Phillips Petroleum Co.)

PIPING SYMBOLS

drawing single line for piping under 12 inches with double line for 12-inch pipe and over.

Because double-line drawings are usually made to scale, it is important for the draftsman to have available information that will enable him to determine the exact allowances that should be made for the specified fittings and valves. The tables in Figs. 2.6 to 2.10, are compiled from manufacturers' catalogs for use in the problems of this book. In actual practice it would be necessary for the draftsman to have available many more tables as well as much additional information pertaining to valves and fittings. The best source of such information is the catalogs of valve manufacturing companies. The draftsman should become thoroughly familiar with the catalog material pertaining to the product he is drawing or designing.

Even though the symbol used may only faintly resemble the part it is intended to represent, the dimensions should be accurate enough that they will present a true picture of the position, the clearance, and the space the part will occupy in the assembly. The draftsman should also take into consideration the

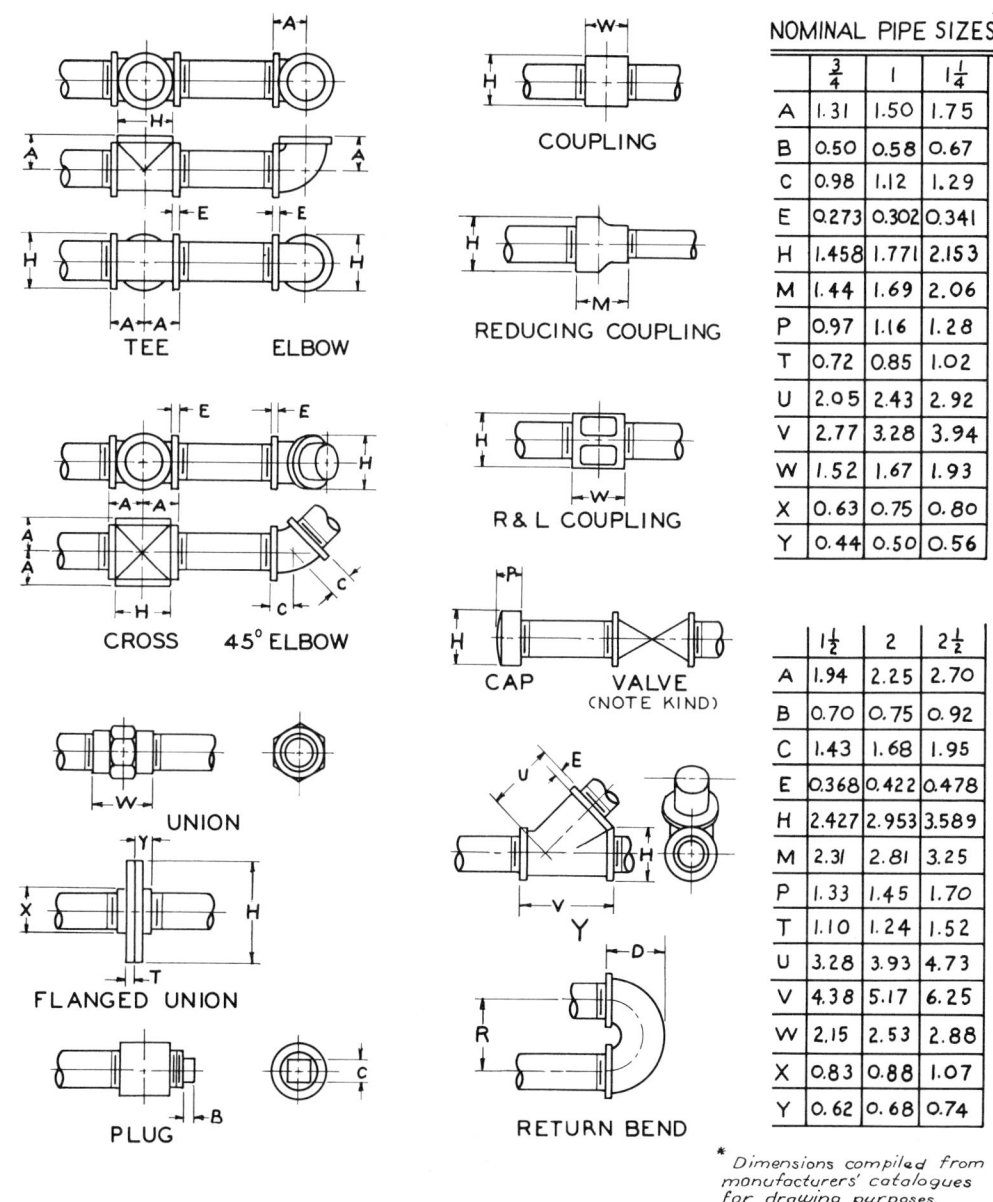

Fig. 2.6. Double-line symbols for screw fittings.

FUNDAMENTALS OF PIPE DRAFTING

space occupied by the gasket in flanged fittings. Since dimensions are usually calculated from center lines it is important that all double-line pipe drawings show the center lines of all pipe. Crossed center lines indicating the centers of valves are also recommended.

POSITION REPRESENTATION

Representation of single-line symbols may be compared to letting a single wire represent the pipe, and representing the connection of the pipe to the fitting by a disc centered on the wire (in the case of flanged fittings, two discs). This assembly when viewed from the top, front, or side will give the appearance of a single-line symbol representing the fitting (Fig. 2.11b).

Most charts refer to the position of the symbols as "in elevation, turned up, or turned down." In elevation means a view looking at the side of the symbol. In the ell, the discs would appear as straight lines.

If the ell is turned so the flow is away from the viewer, one of the connecting discs would appear as a circle and the back of the elbow would appear as a line drawn to the center of the disc (Fig. 2.12).

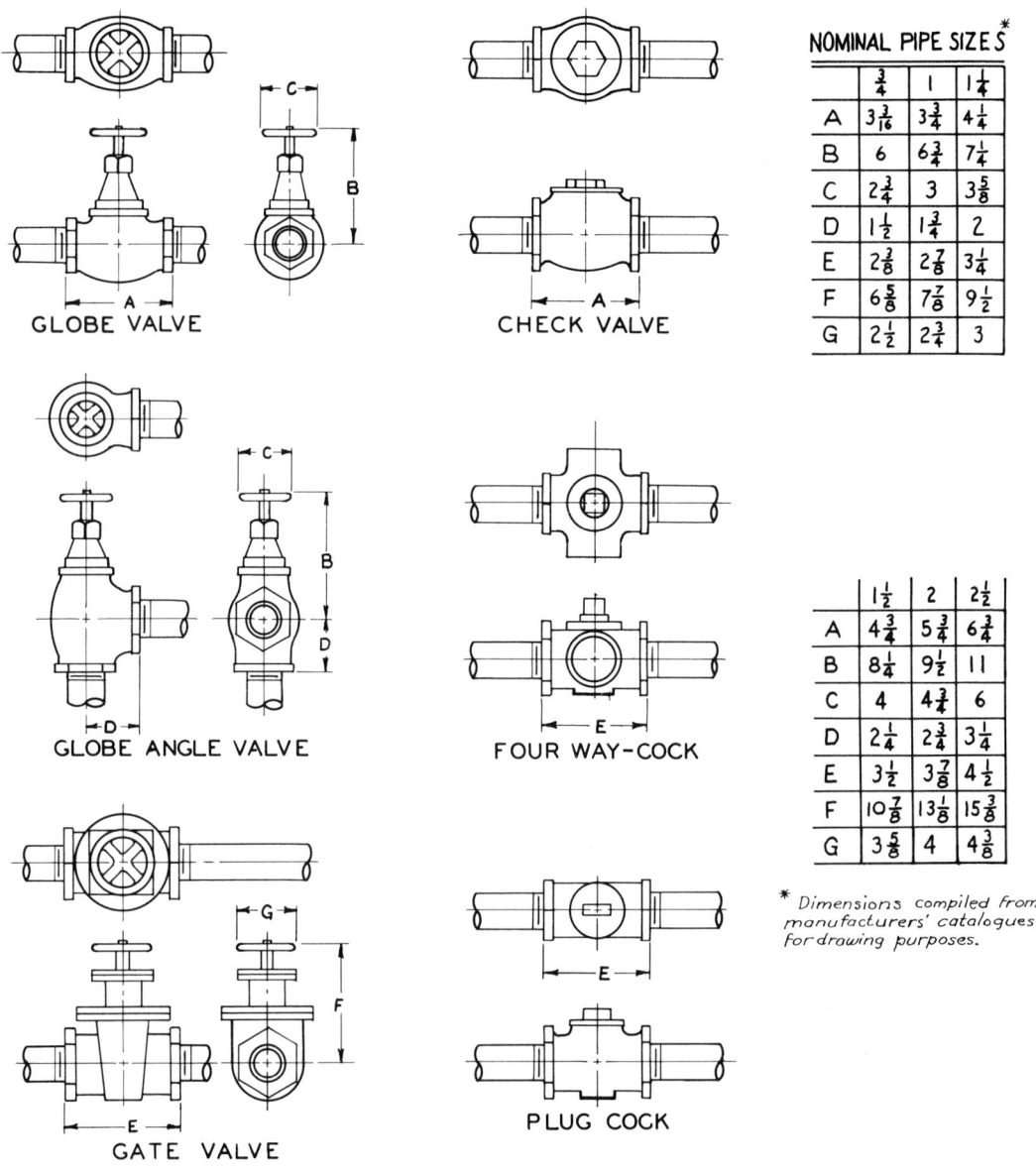

Fig. 2.7. Double-line symbols for screwed valves.

PIPING SYMBOLS

If the elbow is turned so the flow is toward the viewer, one connection would appear as a circle, but the elbow would be hidden from view. The end of the pipe would be shown as a dot in the center of the disc (Fig. 2.13).

DRAWING SYMBOLS

As in all kinds of drafting, a consistent procedure in the development of a drawing or design is one of the greatest assets in the satisfactory execution of the problem. A carefully made freehand drawing should be the first step. The location on the sheet, the placement of details, the arrangement of notes, the scale to be used, and many other problems pertaining to the drawing can be studied before the time-consuming task of putting the drawing on paper in its final form is begun. Once the drawing is carefully thought through and sketched, the process of putting the drawing on paper in its finished form can begin.

In pipe drafting the usual rules relative to the use of equipment should apply. Most drawings are made directly on tracing paper or pencil linen, thus avoiding

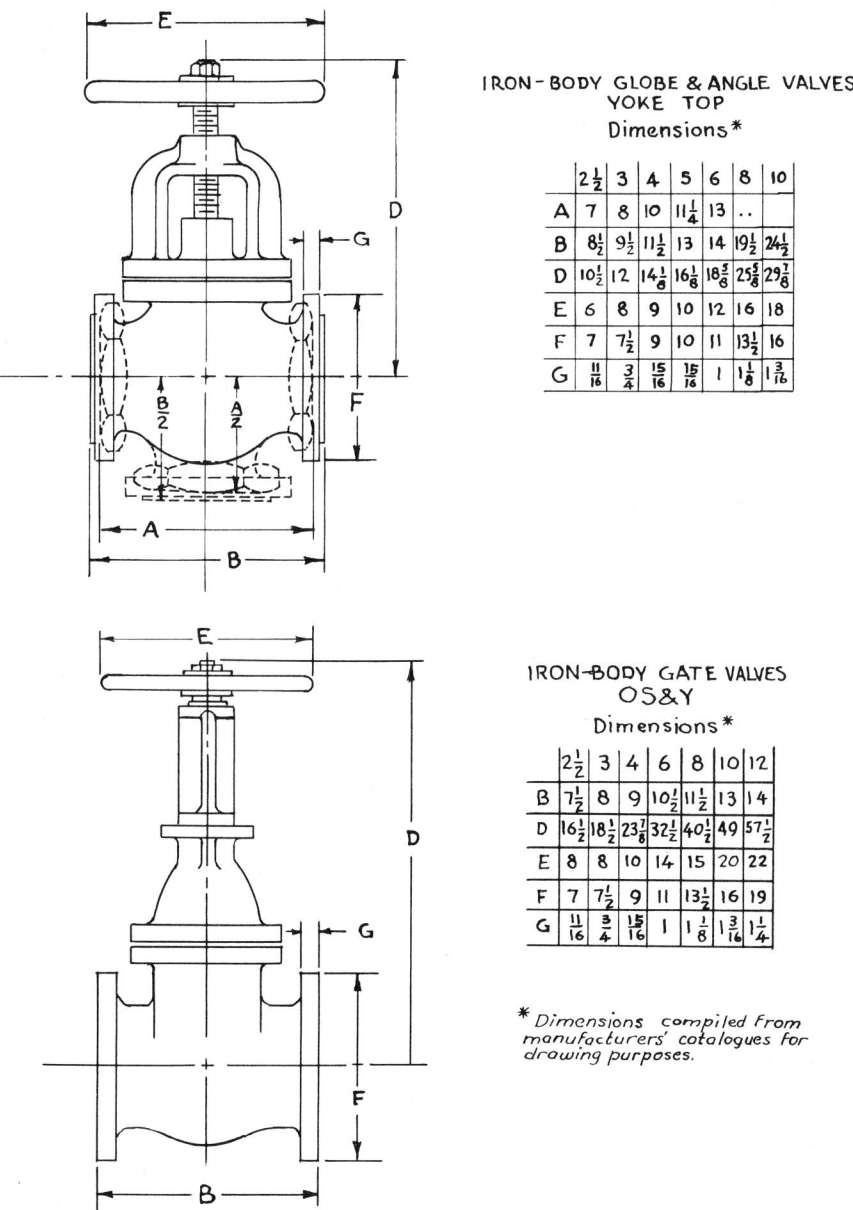

IRON-BODY GLOBE & ANGLE VALVES
YOKE TOP
Dimensions*

	$2\frac{1}{2}$	3	4	5	6	8	10
A	7	8	10	$11\frac{1}{4}$	13	..	
B	$8\frac{1}{2}$	$9\frac{1}{2}$	$11\frac{1}{2}$	13	14	$19\frac{1}{2}$	$24\frac{1}{2}$
D	$10\frac{1}{2}$	12	$14\frac{1}{8}$	$16\frac{1}{8}$	$18\frac{5}{8}$	$25\frac{5}{8}$	$29\frac{7}{8}$
E	6	8	9	10	12	16	18
F	7	$7\frac{1}{2}$	9	10	11	$13\frac{1}{2}$	16
G	$\frac{11}{16}$	$\frac{3}{4}$	$\frac{15}{16}$	$\frac{15}{16}$	1	$1\frac{1}{8}$	$1\frac{3}{16}$

IRON-BODY GATE VALVES
OS&Y
Dimensions*

	$2\frac{1}{2}$	3	4	6	8	10	12
B	$7\frac{1}{2}$	8	9	$10\frac{1}{2}$	$11\frac{1}{2}$	13	14
D	$16\frac{1}{2}$	$18\frac{1}{2}$	$23\frac{7}{8}$	$32\frac{1}{2}$	$40\frac{1}{2}$	49	$57\frac{1}{2}$
E	8	8	10	14	15	20	22
F	7	$7\frac{1}{2}$	9	11	$13\frac{1}{2}$	16	19
G	$\frac{11}{16}$	$\frac{3}{4}$	$\frac{15}{16}$	1	$1\frac{1}{8}$	$1\frac{3}{16}$	$1\frac{1}{4}$

*Dimensions compiled from manufacturers' catalogues for drawing purposes.

Fig. 2.8.

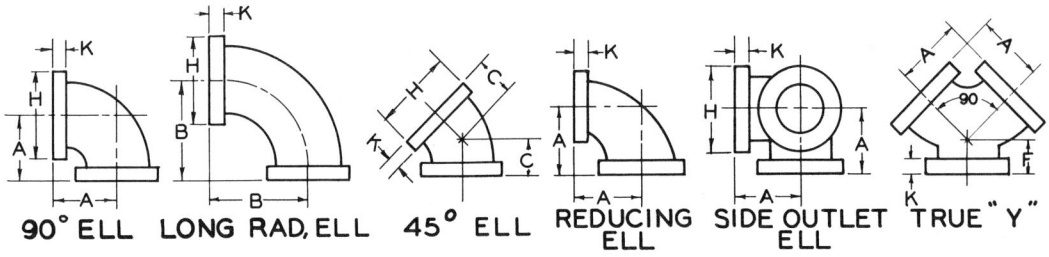

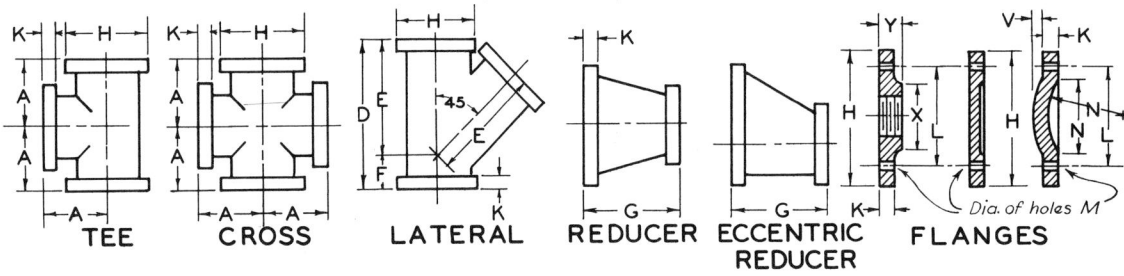

NOMINAL PIPE SIZE N	A	B	C	D	E	F	G	H	K MIN.
	3½	5¾	1¾	7½	5¾	1¾	....	4¼	7/16
¾	3¾	5½	2	8	6¼	1¾	....	4⅝	½
½	4	6	2¼	9	7	2	....	5	9/16
2	4½	6½	2½	10½	8	2½	5	6	5/8
2½	5	7	3	12	9½	2½	5½	7	11/16
3	5½	7¾	3	13	10	3	6	7½	3/4
3½	6	8½	3½	14½	11½	3	6½	8½	13/16
4	6½	9	4	15	12	3	7	9	15/16
5	7½	10¼	4½	17	13½	3½	8	10	15/16
6	8	11½	5	18	14½	3½	9	11	1
8	9	14	5½	22	17½	4½	11	13½	1⅛
10	11	16½	6½	25½	20½	5	12	16	1 3/16
12	12	19	7½	30	24½	5½	14	19	1¼

NOMINAL PIPE SIZE N	L	M	NUMBER OF BOLTS	DIA. OF BOLTS	LENGTH OF BOLTS	X MIN	Y MIN	WALL THICKNESS	V
1	3⅛	5/8	4	½	1¾	1 15/16	11/16	5/16	3/8
1¼	3½	5/8	4	½	2	2 5/16	13/16	5/16	7/16
1½	3 7/8	5/8	4	½	2	2 9/16	7/8	5/16	½
2	4¾	3/4	4	5/8	2¼	3 1/16	1	5/16	5/8
2½	5½	3/4	4	5/8	2½	3 9/16	1⅛	5/16	5/8
3	6	3/4	4	5/8	2½	4¼	1 3/16	3/8	11/16
3½	7	3/4	8	5/8	2¾	4 13/16	1¼	7/16	3/4
4	7½	3/4	8	5/8	3	5 5/16	1 5/16	½	7/8
5	8½	7/8	8	3/4	3	6 7/16	1 7/16	½	7/8
6	9½	7/8	8	3/4	3¼	7 9/16	1 9/16	9/16	15/16
8	11¾	7/8	8	3/4	3½	9 11/16	1¾	5/8	1 1/16
10	14¼	1	12	7/8	3¾	11 15/16	1 15/16	3/4	1⅛
12	17	1	12	7/8	3¾	14 1/16	2 3/16	13/16	

Fig. 2.9. American Standard cast-iron flanges and flanged fittings, 125 psi. (Data from ASA B 16a–1939.)

PIPING SYMBOLS

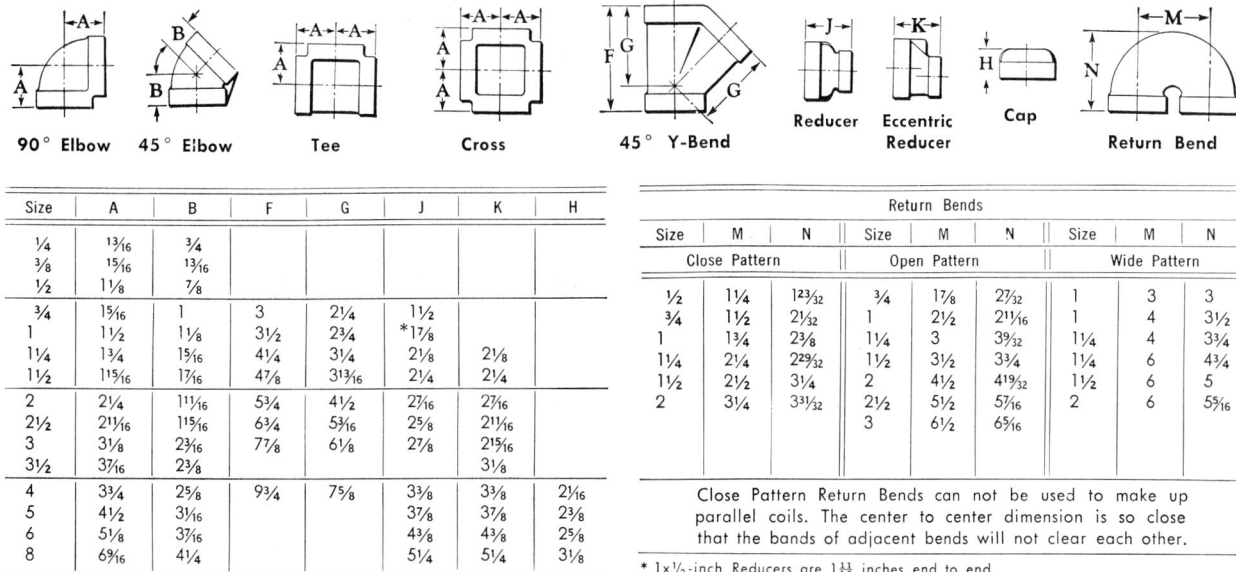

Fig. 2.10a. Standard cast-iron fittings. Dimensions in inches. (Reproduced by permission of the Crane Co.)

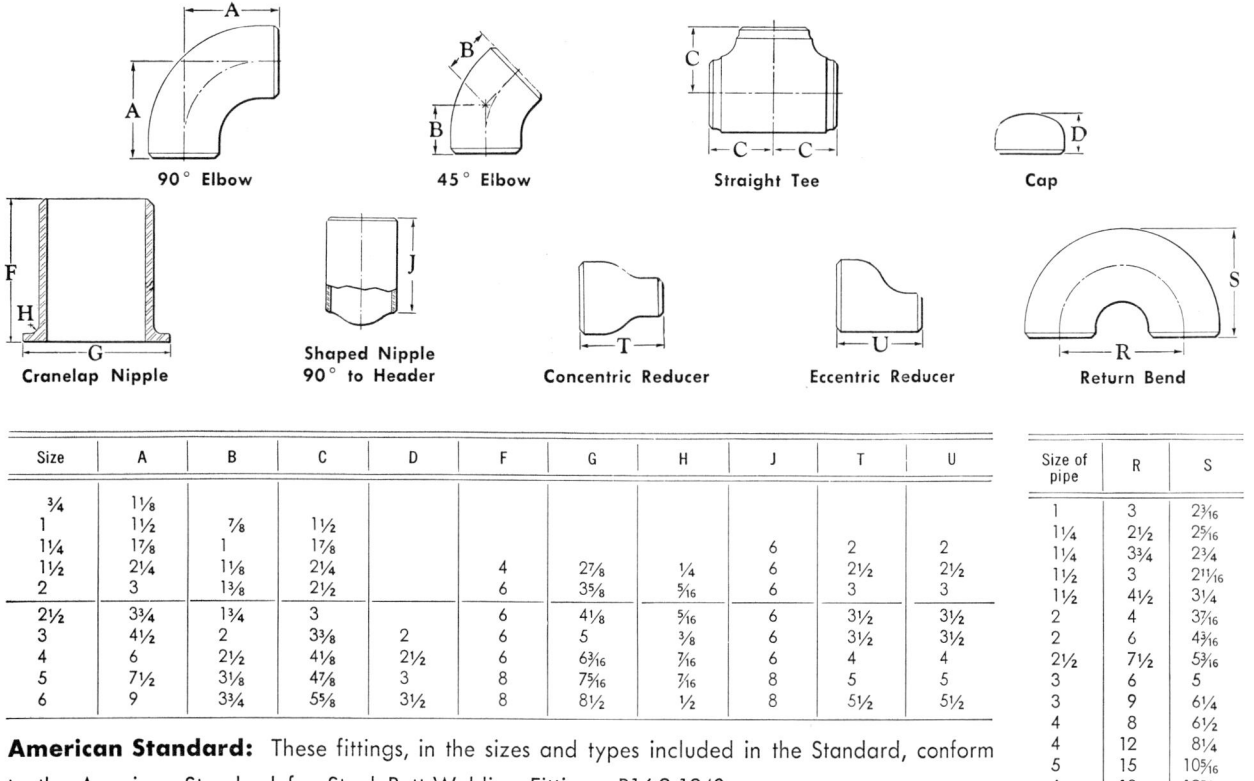

American Standard: These fittings, in the sizes and types included in the Standard, conform to the American Standard for Steel Butt-Welding Fittings, B16.9-1940.

Fig. 2.10b. Steel butt-welding fittings for use with standard pipe. Dimensions in inches. (Reproduced by permission of the Crane Co.)

FUNDAMENTALS OF PIPE DRAFTING

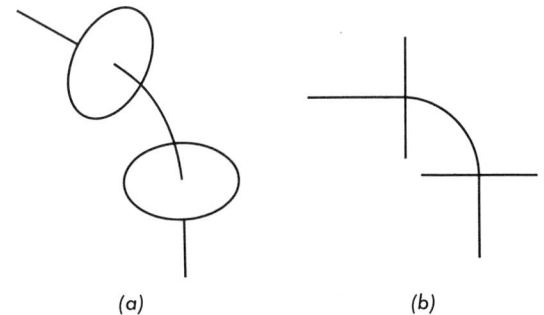

Fig. 2.11. (a) Perspective of disc arrangement on wire to represent a screwed ell. (b) Side view of disc arrangement on wire showing how the single-line symbol for a screwed ell is evolved.

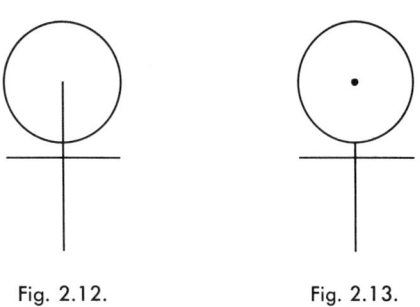

Fig. 2.12.   Fig. 2.13.

It should be pointed out that almost every piping layout will contain parts for which no symbols exist. In the absence of a standard symbol, the part in question should be outlined to resemble the appearance of the part or machine and be accompanied by an explanatory note.

After all symbols are laid out and all lettering has been completed, the light construction lines which show the outlines of the symbol should be traced over with dark, firm lines approximately $\frac{1}{32}$ inch in width.

the intermediate step of tracing. If extra quality is desired, ink on tracing cloth is recommended.

The first step in placing the drawing on tracing paper or linen is to lay out the center lines for all pipe in a given circuit or system. These layout lines should be very fine, sharp lines made with a hard pencil. This procedure applies in making both single- and double-line symbols.

The second step is to lay out the symbols in their correct locations (the center of each symbol should be located by a light construction line crossing the center line of the pipe) and in their correct sizes and proportions. This should be done by laying out first the arcs for the elbows and then the circles representing the end views of the fittings and valves. By projecting from the outside diameters of the end circles the lengths of the crosses representing the ends of the fittings and valves can next be drawn. Care should be taken to see that these end crosses are properly spaced by measuring along the center lines of the pipe.

After the ends of the symbols are drawn, the crosses representing the valves should be made. Due to the large number of possible symbols, it is inadvisable to discuss every detail of every one. Close observation of the symbol chart should enable the student to draw readable symbols.

Problems

General Instructions

1. Sheet layout. All drawings, unless otherwise stated, are to be laid out in compliance with the diagram of Fig. 2.14.

The title block is to be laid out in the lower right-hand corner of the drawing area on all sheets, which are to be divided as indicated in Fig. 2.15.

All lettering in the title block is to be vertical upper case $\frac{1}{8}''$ high, with the exception of the drawing title which is to be $\frac{3}{16}''$ high.

2. Lettering.

A. Names of parts, when appearing outside of the title block, should be inclined upper-case letters.
B. Names of parts appearing in the material bill should be inclined lower-case letters.
C. Each word in notes and instructions should begin with an inclined upper-case letter, the rest of the word inclined lower case. *Exception:* The words "a," "an," and "the" used before nouns to limit their application should begin with lower-case letters except at the beginning of a sentence.

Many draftsmen prefer to use upper-case letters exclusively, preferably vertical; however, many companies still prefer inclined lower-case. For this reason, it is recommended that the student become adept in the use of both types of lettering.

3. Drawing material.

A. Equipment:
  1. Drawing set
  2. 30°–60° triangle
  3. 45° triangle
  4. T-square
  5. Erasing shield
  6. Architect's scale
B. Supplies:
  1. 20 sheets 12″ x 18″ tracing paper
  2. 4 sheets 24″ x 36″ tracing paper
  3. 4 sheets 24″ x 36″ white detail paper for backing
  4. 2H pencil

PIPING SYMBOLS

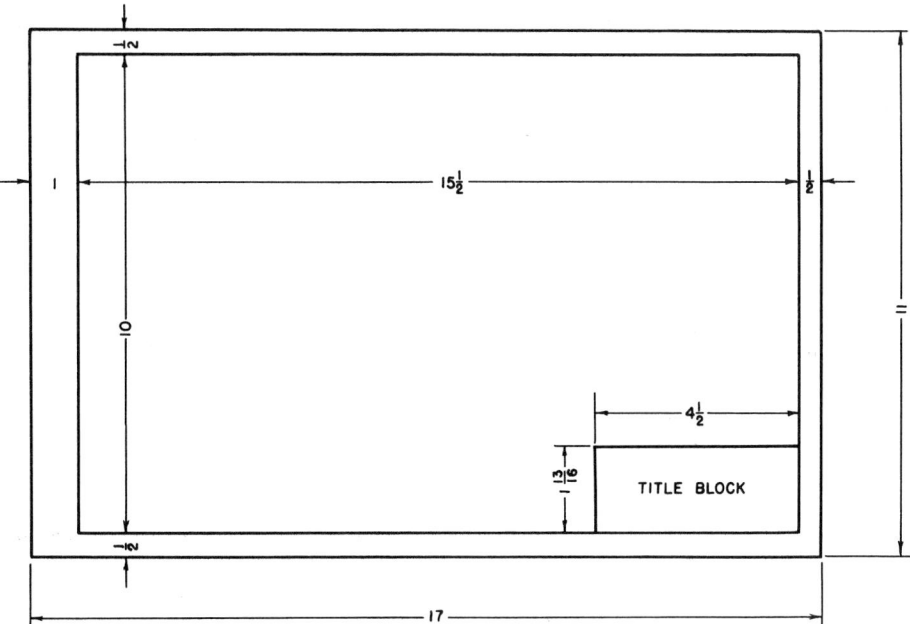

Fig. 2.14.

5. 4H pencil
6. Ruby Eberhard eraser
7. Pink Pearl eraser
8. Roll of drafting tape

4. Finishing. All problems are to be finished in pencil on tracing paper unless otherwise specified.

Problem 1. Draw the piping layout of Fig. 2.16 on a sheet laid out as outlined in the general instructions.

The numbers in the circles of Fig. 2.16 are those accompanying the symbols in the symbol charts in Fig. 2.1 to 2.4.

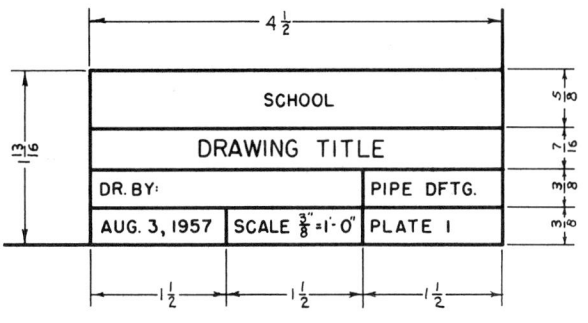

Fig. 2.15.

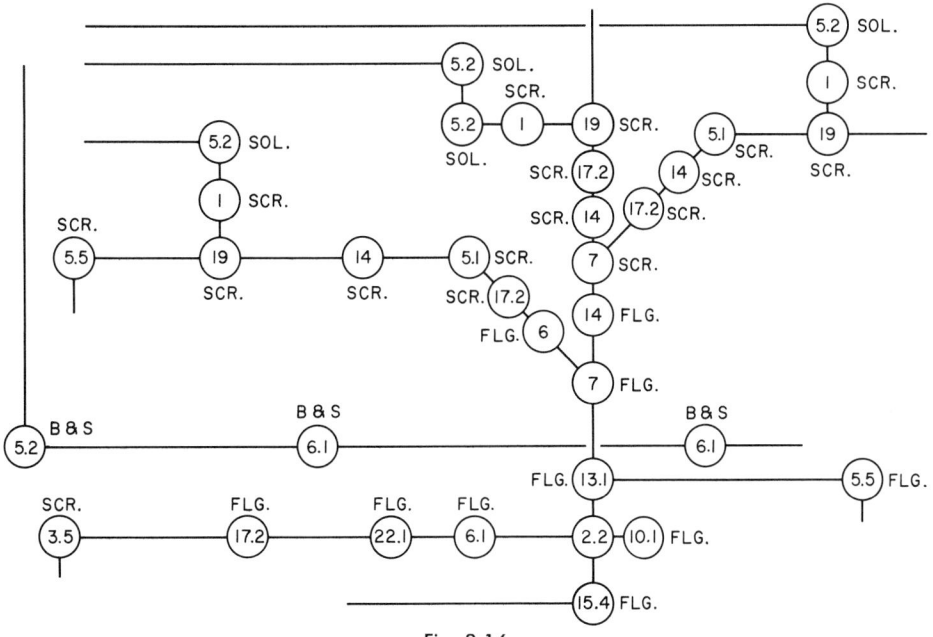

Fig. 2.16.

Replace the circles and letters with the appropriate symbols; also letter the names of the fittings or valves in an area adjacent to the part named.

Arrange the drawing to make a good appearance on the sheet.

Title the sheet "PIPING LAYOUT."

Problem 2. Make a double-line drawing of the single-line layout shown in Fig. 2.17. Use all flanged valves and fittings. In order to obtain proportions of valves and fittings, assume all pipe to be 6″ and draw to an appropriate scale. Refer to table in Fig. 6.2 for actual nominal pipe sizes. Proportions of valves and flanged fittings to be used may be found in Figs. 2.8 and 2.9. Use the simplified piping symbols illustrated in Fig. 2.5. These tables are for problem reference only. More complete tables can be found in valve catalogs.

Draw on the same sheet size used in Problem 1, and arrange for a good appearance. Identify each part by note as well as by symbol.

Title the sheet "PIPING LAYOUT."

Problem 3. Follow the instructions of Problems 1 and 2 and draw Fig. 2.18 in double line using all screwed fittings. The tables accompanying Figs. 2.6, 2.7, and 2.10a may be used to obtain the required dimensions.

Assume main pipe to be 2″ in diameter. Also assume that your drawing is to be used for display; therefore, all valves and fittings should be drawn to give a realistic appearance. Scale is important only in that it regulates the proportions of the parts.

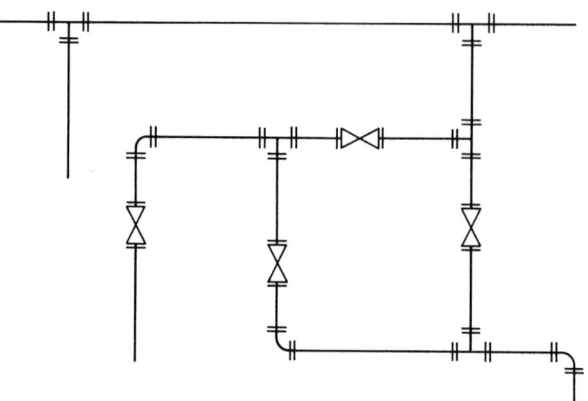

Fig. 2.17.

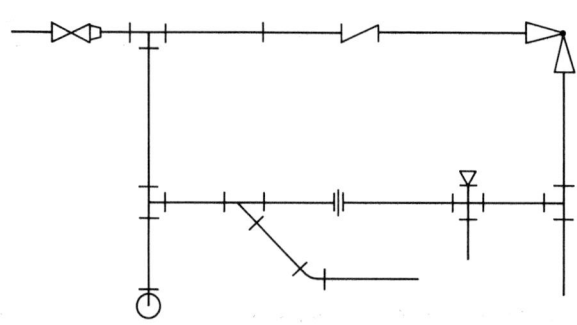

Fig. 2.18.

3 Methods of representing piping systems

A piping system may be presented by the following three methods: (1) orthographic projection, (2) developed-view method, and (3) pictorial representation.

ORTHOGRAPHIC PROJECTION

The method of representation with which the machine draftsman is most familiar is that of orthographic projection. This consists of a top or plan view, an elevation or front view, and a side or profile view. Many piping drawings may consist of only one of the views. This would be the case if most of the pipe is to be installed in one plane, either horizontal or vertical, or if the draftsman is mainly interested in showing the relation of the piping to the parts or machines it connects, leaving the details of the installation to the worker.

If complete and accurate details for a small portion of an installation are desired, the orthographic method of representation offers the best means for supplying the information. One, two, three, or more views may be utilized. Details of simple assemblies or parts can best be shown by this method because it is very adaptable to dimensioning. Piping layouts drawn on building plans to show the relation of the piping to the building features are drawn orthographically.

The main objection to the use of orthographic projection is the difficulty of interpreting information if the system consists of a number of overlapping pipes, valves, and fittings. An orthographic projection of an intricate piping system could easily become so complicated as to defy the skill of an experienced blueprint reader. When this occurs, it is common practice to accompany the orthographic projection with a pictorial diagram. Pictorial drawing will be discussed later in this chapter.

An example of a simple pipe assembly drawn orthographically is shown in Fig. 3.1.

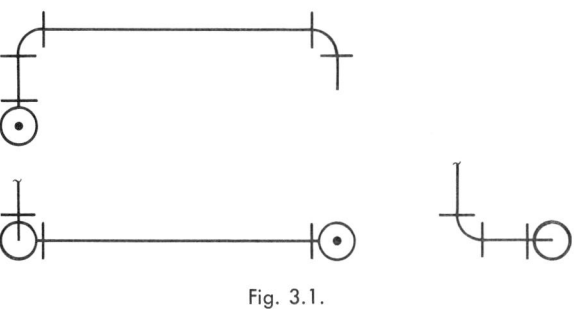

Fig. 3.1.

DEVELOPED VIEWS

Situations often arise in which the conventional application of orthographic projection is not advisable because of the resulting confusion of details. In order to separate and identify the parts, the system of the developed view is used. This consists simply of imagining the entire system to be flattened into one plane, either vertical or horizontal. All fittings would then be turned sideways and all pipe would appear in its true length without the confusing effect of crossed and overlapping pipes, valves, and fittings. This procedure simplifies for the workman the identification of the separate parts of the system and facilitates the compilation of the parts list and material bill. There is a close relation between developed views and diagrams showing the working principles of a system.

Figure 3.2 shows a developed view of the system shown orthographically in Fig. 3.1.

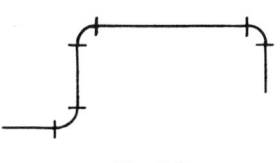

Fig. 3.2.

PICTORIAL REPRESENTATION

The third and most easily interpreted kind of piping drawings are made by some method of pictorial representation. The method may be oblique drawing, isometric drawing, or axonometric drawing. Some piping draftsmen use various methods of fake perspective not common to the usual methods of pictorial drawing.

Oblique drawing. In oblique drawing the piping in one plane of the drawing is drawn as a true projection. The plan view is most frequently used for this purpose. All piping from this plane is drawn obliquely. Figure 3.3 represents a one-pipe heating system drawn obliquely. It gives the effect that the reader is standing well above the center of the plan view and that all pipe that rises above the plan view is extending toward the plane in which the reader is standing. The system, illustrated orthographically in Fig. 3.1 and as a developed view in Fig. 3.2, is shown as an oblique drawing in Fig. 3.4.

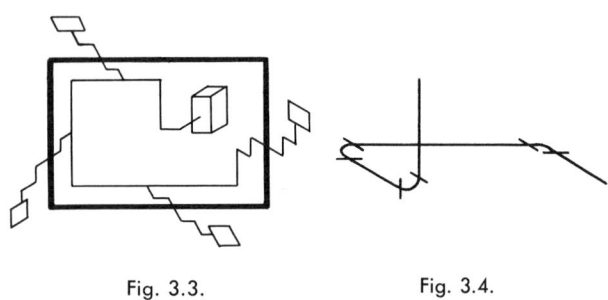

Fig. 3.3.   Fig. 3.4.

Isometric drawing. The drawing of piping in isometric follows the conventional rules of isometric drawing in that all vertical lines are drawn vertical, but lines that recede to the right and to the left are drawn at an angle of 30° to the horizontal.

The piping system illustrated in Figs. 3.1 to 3.4 is drawn isometrically in Fig. 3.5. Note that the cross

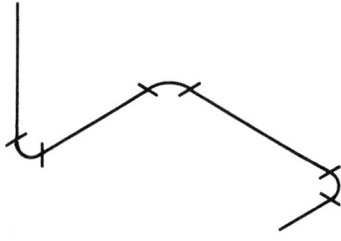

Fig. 3.5.

lines on the symbols are drawn so they appear to be lying in the plane formed by the bend in the pipe.

Axonometric drawing. This is a method of pictorial drawing often used by piping draftsmen. It has an advantage over isometric drawing in that it is possible to crowd more details into a rectangular space, thus better fitting the area of the average drawing sheet. If a proper axis is selected, it also gives a more natural appearance to the piping in the system.

Any draftsman who is capable of drawing in isometric should also be able to draw axonometrically. Although the three axes of an isometric drawing must form three angles of 120° each, the axes of an axonometric drawing may form any preselected angles with each other. Figure 3.6 compares the only axis arrangement of the isometric with one of the innumerable axis arrangements of the axonometric.

When using the axonometric axis illustrated, a 15° triangle would be useful, otherwise it is necessary to use the 30°-60° and the 45° triangles in combination to obtain the required 15° for the two receding axes. The receding lines are foreshortened by three-fourths of their actual scaled length to give a more realistic appearance. If double-line drawing is used, the formation of ellipses representing circles presents somewhat of a problem unless an appropriate ellipse template is available.

When drawing by any of the pictorial methods mentioned, considerable skill and time is involved, but in many situations the improved appearance and readability of the drawing justifies the effort.

Problems

Problem 1. Draw a double-line developed view of the oblique piping assembly in Fig. 3.7. Use the double-line symbols illustrated in Fig. 2.6.

Use a scale of ⅜″ = 1″. Draw 1″ pipe.

Center-to-center distances are not important. Arrange for a balanced appearance on an 11 x 17 sheet.

METHODS OF REPRESENTING PIPING SYSTEMS

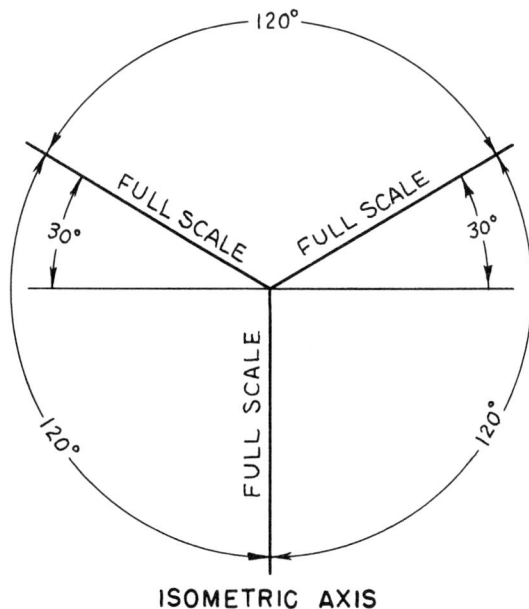
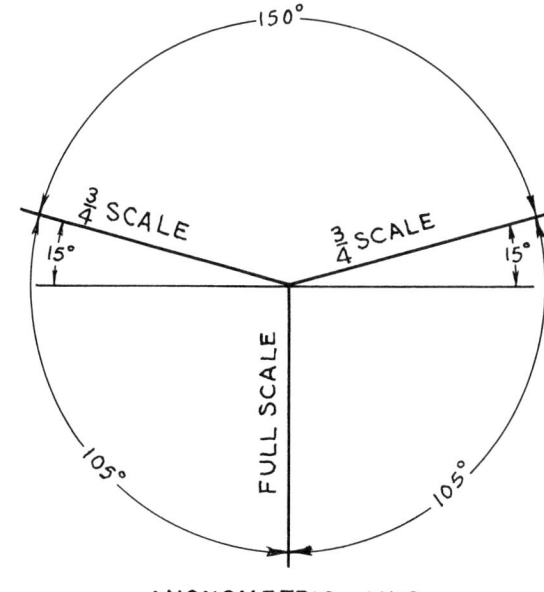

Fig. 3.6.

Problem 2. Figure 3.8 shows three views of a small portion of the piping system for a high-pressure condensate flash tank.

Redraw in double-line drawing the plan and elevation described by Fig. 3.8. The pipe size should be 1″.

Select a scale appropriate to an 11 x 17 sheet.

On the same sheet draw a small single-line isometric diagram showing valves and fittings. Indicate flow by arrows.

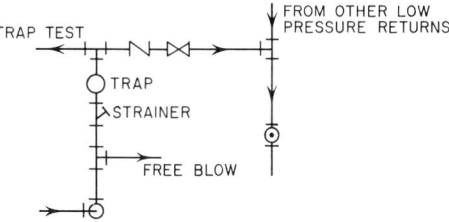

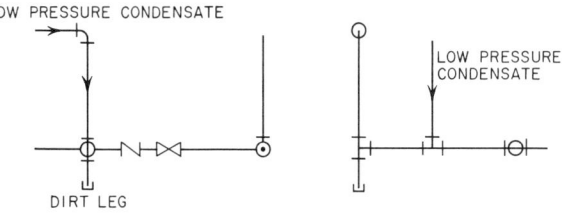

NOTE:
IN THE EVENT OF OVERLAPPING SYMBOLS, THE REAR OR OVERLAPPED SYMBOL IS USUALLY OMITTED.

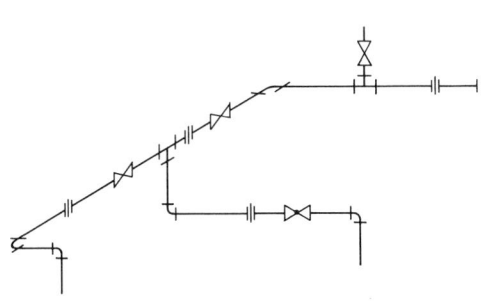

Fig. 3.7.

Fig. 3.8.

4 Diagram drawing

The piping draftsman, if he is to advance in his work, must be able not only to draw a piping system, but also to understand the operating principles. He should be able to analyze complicated piping arrangements into easily understood diagrams, or conversely, to turn a diagram into a working system with all the necessary construction details included. The draftsman should be able to use data arrived at by the engineer, and to use this data with common sense, and in a practical way. All the hundreds of details established by common practice should be the responsibility of the draftsman, leaving the design engineer free for problems of an engineering nature.

Piping systems for established situations are usually very similar in their application; therefore, if the draftsman is able to understand the common and accepted systems, he can relieve the engineer of routine jobs and at the same time make for himself a position of responsibility in his company.

By understanding the workings of various typical systems the draftsman will also be enabled to more easily understand variations of those systems. Practice in interpreting the operating principles of a system of piping is comparable to practice in blueprint reading for the machine draftsman. There is no substitute for practice, and no magic formula for the development of the ability to read piping drawings. When the process of operation is understood, the draftsman is in a position to make decisions that will lead to the efficient design of the system.

The diagram is the piping draftsman's shorthand, and he should understand its place in the planning of a piping system.

SINGLE-LINE DIAGRAM DRAWINGS

Single-line diagram drawings are usually made for the purpose of simplifying a piping system. Actually, all single-line drawings are diagram drawings, even when they give all information pertinent to the construction of a piping system.

SCHEMATIC DRAWINGS

The term "schematic drawing" usually refers to the drawing of the theoretical layout of a piping system. Such a drawing can be used by the draftsman as a guide for the layout of piping in a building by using the principles illustrated and adapting them to a given situation. This type of drawing is often used in advertising literature and in text books. It may vary from the simplest single-line diagrams to colored and shaded illustrations.

FLOW DIAGRAMS

Flow diagrams are a type of schematic drawing. In the simplest form, they are single-line drawings in which equipment is represented by simple geometric figures, and in which the arrangement of the equipment is only general. Piping is represented by single lines and the direction of flow of fluids or gases in the pipe is indicated by arrows. A flow diagram may also be used as a basis for the further study of a system, for the analysis of heat loads, or

DIAGRAM DRAWING

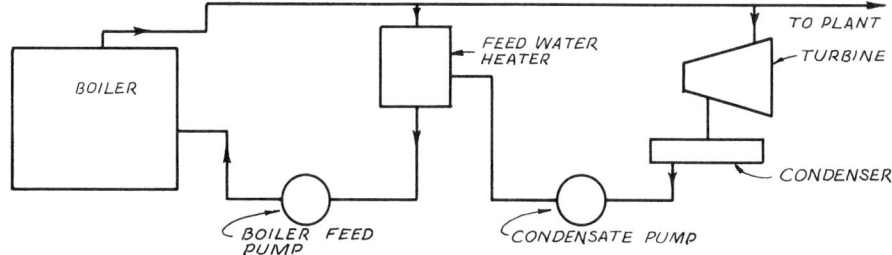

Fig. 4.1. Simple steam-flow diagram.

for the sizing of pipes. This type of drawing is usually the first step in the designing of a system. Figure 4.1 illustrates a simple steam flow diagram. It shows the steam flow from a boiler to a turbine and to the plant.

Frequently separate diagrams can be combined into one. The drawing of Fig. 4.1 could have been enlarged to show the flow of the steam to each piece of equipment in the plant, and a dotted line could be used to show the condensate flow from the equipment back to the boiler.

Problems

Problem 1. Figure 4.2 is a drawing of the lube oil filter piping for a five-cylinder reciprocating pump. The drawing shows the piping hookup as well as the sizes of the pipe and fittings. No center-to-center dimensions are given.

Redraw on an 11 x 17 sheet. Use symbols and symbol sizes recommended in Figs. 2.1 to 2.4. Draw the lube oil pump and filter to approximately twice the size illustrated.

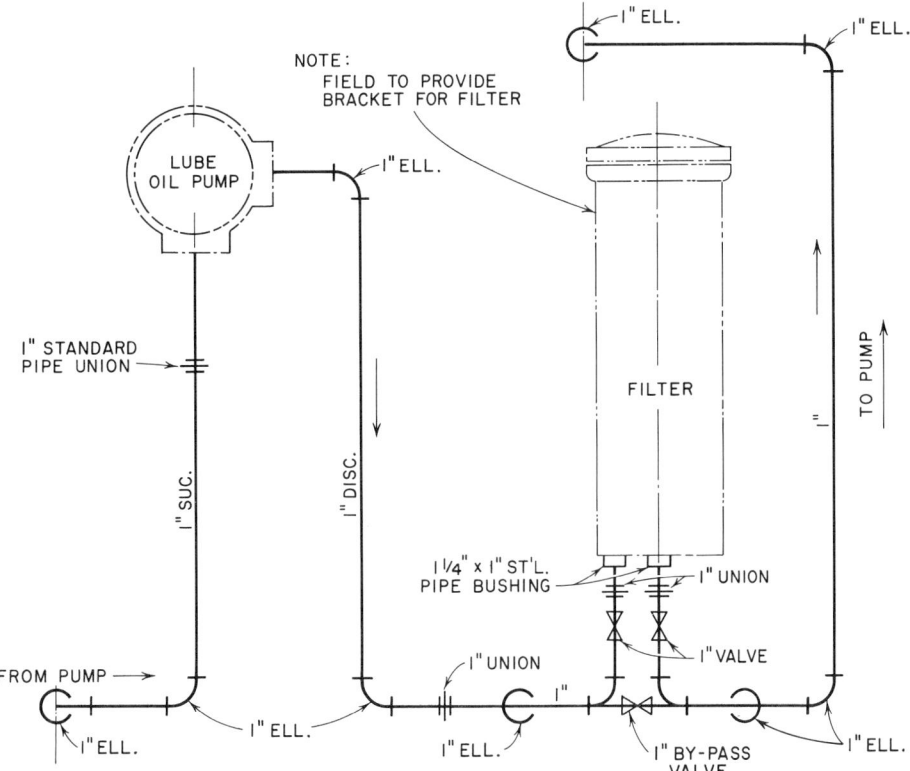

Fig. 4.2. Lube oil filter for five-cylinder reciprocating pump.

FUNDAMENTALS OF PIPE DRAFTING

Arrange the drawing on the sheet so as to allow space in the upper right-hand corner for an isometric schematic sketch of the filter and piping to the filter including the by-pass. Do not show the lube oil pump on the schematic.

Label all fittings on the isometric with the correct name and size.

The piping from the lube oil pump to the filter is so arranged as to pass from the lube oil pump to the bottom of the filter. After being filtered, the oil passes out the bottom of the pump. A by-pass is so arranged that the valves to the filter may be closed and the by-pass valve opened allowing oil to pass from the lube oil pump to the five-cylinder reciprocating pump without passing through the filter. Note that all necessary fittings have not been shown, nor has the draftsman complied with American Standard recommendations for symbols. For ells that turn away from the viewer, a broken circle was used with the opening toward the direction in which one leg of the ell points.

Problem 2. Figure 4.3 is a diagrammatic drawing of detail "A" taken from a larger print showing the service piping for fuel oil and gas, starting air, and water of a pump station. Notice that this detail shows only piping and tanks pertaining to the starting air.

Study the operating principles and redraw as a simple, nonisometric schematic drawing without regard to pipe sizing. Use a single line and disregard all fittings. Use arrows to show direction of flow. Show all valves.

All piping and tanks should be drawn in a single plane with a minimum of overlapping and crossing.

Draw on an 11 x 17 sheet.

Label valves and lines as required.

Title the sheet "AIR SUPPLY SYSTEM."

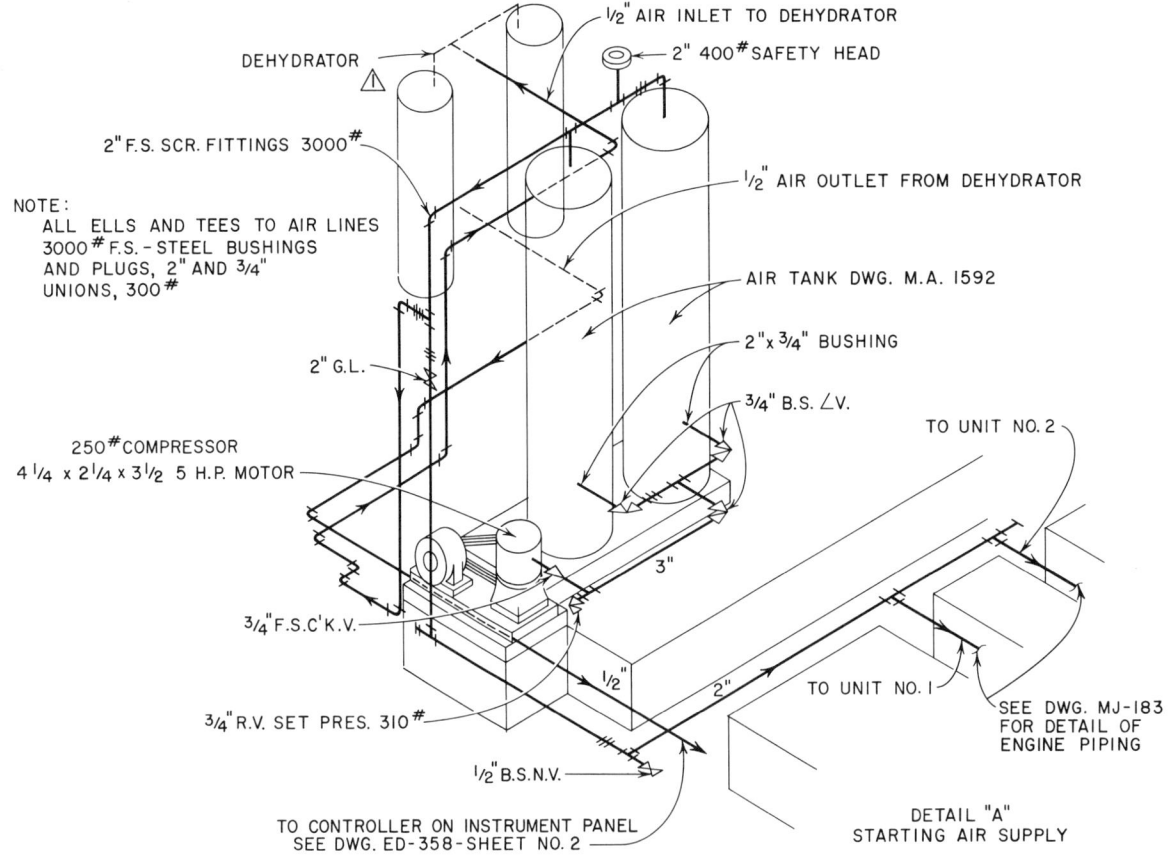

Fig. 4.3.

5 Controls

The purpose of any piping system is to direct and control the flow of liquids or gases, which may vary from intense cold to extreme heat, from extremely high pressures to low pressures, or from extremely viscous to extremely volatile. It is usually the responsibility of the client or the engineer piping designer to select correct valves or controls for a given situation, however, it is advisable that the draftsman understand, in addition to the symbolic representation of controls, the principles of their operation and their application to given conditions.

VALVE DESIGNS

Valves may be divided into seven basic designs as follows: (1) gate, (2) globe, (3) relief, (4) angle, (5) cock, (6) swing check, (7) lift check. These basic valve designs are illustrated in Fig. 5.1.

The five principal functions of valves are: (1) start and stop flow, (2) regulate or throttle flow, (3) prevent back flow, (4) regulate pressure, (5) relieve pressure.

The operating principles of some of the basic valve designs, each of which has many variations, are discussed below.

Globe valves. The globe valve, as illustrated in Fig. 5.1a, is designed to cause a change of direction of the liquid flowing through the valve, thereby creating a resistance to the flow and permitting close regulation of flow. The design is usually such that the seat and disc can be quickly reseated or replaced, making them ideal for services that require frequent maintenance. Angle globe valves have the same characteristics as the straight globe valves except that the inlet and outlet are set at right angles to each other which, in many situations, may add to the convenience of operation, flexibility of placement, and at the same time eliminate the use of an elbow. The disc of the globe valve may be flat with a composition face like the one illustrated in Fig. 5.3a, it may be a plug type as the one illustrated in the check valve in Fig. 5.1e, or it may be the so-called conventional type illustrated in Fig. 5.1a which has a chamfer on the disc which forms a narrow line of contact with the seat.

The flat composition disc globe valve has the advantage of adaptability to different services because the composition disc can be changed to resist the action of the fluid being conveyed. Composition discs are suitable to all moderate-pressure services except throttling. They will often stand imbedding of dirt without leaking.

The plug-type disc, due to the long taper on the plug and the matching seat, offers high resistance to the cutting effect of dirt and other foreign matter. It is adaptable to such services as throttling, drip and drain lines, soot blowers, blowoff, and boiler feed lines.

In contrast to the plug-type disc, the conventional-type disc has a thin line of contact which serves to break down hard deposits that form on the seats, thus assuring pressure-tight closure. The conventional disc-type valve is widely used for many hot and cold services.

When reduced flow or pressure drop caused by the change of direction of flow within the valve is not objectionable, and when throttling and flow regulation are desirable, the use of the globe valve may be justified. The smaller sizes are considered suitable for most services, but the use of this type of valve in sizes

FIG. 5.1. BASIC VALVE DESIGNS

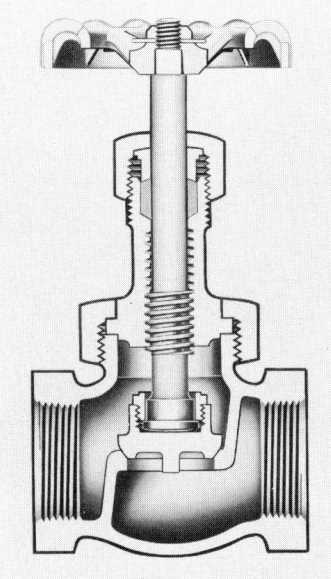

Fig. 5.1a. Globe valve. (Courtesy Crane Co.)

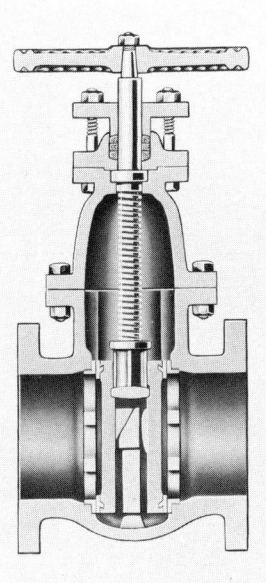

Fig. 5.1b. Double-disc gate valve. (Courtesy American Chain and Cable Co., Inc.)

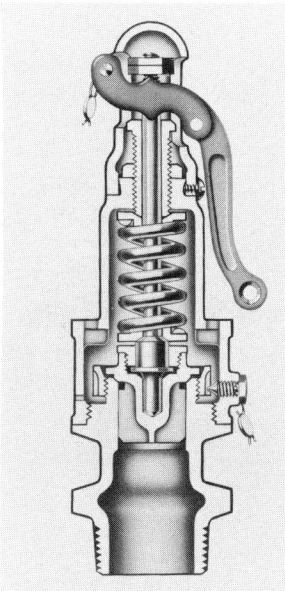

Fig. 5.1c. Relief valve. (Courtesy Crane Co.)

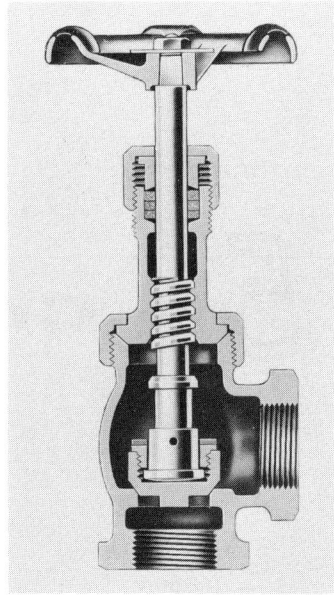

Fig. 5.1d. Angle valve. (Courtesy American Chain and Cable Co., Inc.)

larger than 6 inches is questioned by most authorities from the standpoint of both efficiency and expense. The cost of globe valves is usually equal to or greater than the cost of either plug or gate valves. This difference in cost increases as the sizes become larger.

Gate valves. Gate valves, one of which is illustrated in Fig. 5.1b, are primarily free-flow valves. When the gate is lifted a free, unobstructed flow passes through the valve. It is not intended as a throttling device. Gate valves are intended for use when the operation is infrequent. The most common type of gate valve is the one in which the gate is a solid wedge disc. The design is simple and the valve may be installed in any position. Its most common use is for steam service.

The double-disc-type gate valve, illustrated in Fig. 5.1b, is one in which the disc passes between parallel or tapered seats in the valve body. In the parallel seat pattern, as the disc reaches the closed position the discs are spread against the body seats. These valves are used widely on noncondensing gas and liquid services at normal temperatures, like waterworks and sewage fields, and in the oil and gas industries. For the most satisfactory services, the parallel disc valves should be installed in an upright position.

Cock valves. The plug, or cock-type valve, also called the plug gate, has many variations. The basic design is that of a cone-shaped plug which serves as the closure. A hole in the plug allows the line to flow when the plug is turned to the open position. This principle is used in the cross valve when it is necessary to change the direction of the flow at the valve. The plug-type valve has many of the advantages of the gate valve, and in addition, it is suited for throttling service; however, as with all valves, the recommendation of the manufacturer should be carefully studied before selection for installation. Figure 5.1f illustrates a cock-type valve.

Some types of plug valves depend for a seal upon lubricant forced into grooves under high pressure. These valves are called "lubricated plug valves." Other valves of the plug type depend entirely upon metal-to-metal contact for seal. This type is known as "dry plug valves."

Check valves. Check valves are nonreturn valves used to prevent back flow in lines. They conform to two basic patterns, the lift check, illustrated in Fig. 5.1e, and the swing check, illustrated in Fig. 5.1g. In the swing check, flow moves through the valve in practically a straight line, and in the lift check, the flow changes course as in a globe valve. In both types, flow keeps the valve open and the reversal of flow closes it. Since both are operated by gravity, they should be installed in a horizontal, upright position. There are, however, check valves designed for mounting in vertical pipe lines. Classified with the check

FIG. 5.1. BASIC VALVE DESIGNS

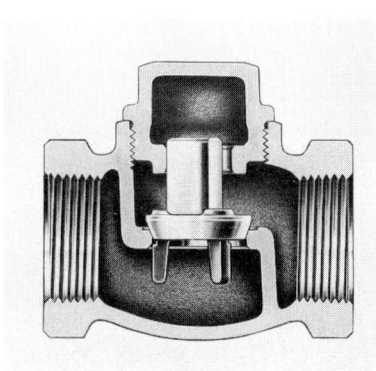

Fig. 5.1e. Lift-check valve.
(Courtesy Lunkenheimer Co.)

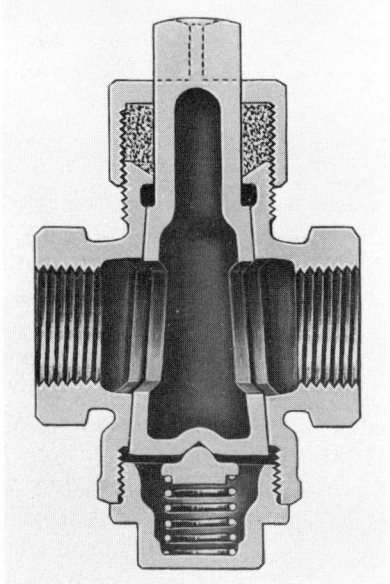

Fig. 5.1f. Cock valve.
(Courtesy Lunkenheimer Co.)

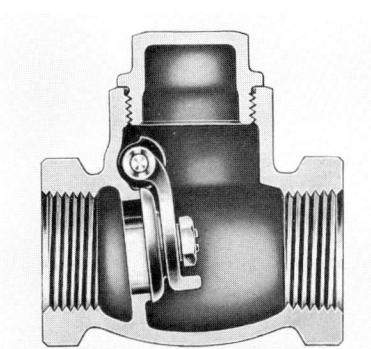

Fig. 5.1g. Swing-check valve.
(Courtesy Lunkenheimer Co.)

valve could also be the automatic stop-check illustrated in Fig. 5.4m. This valve is actually a check valve with a hand control. In its open position, this valve will permit flow in only one direction.

Relief valves. The relief valve illustrated in Fig. 5.1c is one of many patterns used for relieving pressure on boilers and other equipment to prevent damage from over pressure. This valve is spring loaded and may be adjusted to relieve pressure at a given point. Relief valves are also called safety valves, the distinction being that safety valves are usually used for steam or other gases, and relief valves are usually used for liquids.

STEM VARIATIONS

In addition to the basic valve design, there are many variations in the stem design of the gate, globe, and angle valves. Figure 5.2 illustrates four of these variations. Figures 5.2b and 5.2d are of the outside screw and yoke type (OS&Y). In Fig. 5.2b the handwheel remains in position and the stem rises. This is called an OS&Y with a rising stem. In Fig. 5.2d the handwheel rises with the stem. In both Figs. 5.2b and 5.2d, since the stem screw remains outside of the body at all times, the threads are not subjected to the effects of fluids in the line; also, the threads are easily accessible for lubrication. The position of the stem also offers visible evidence as to the open or closed position of the valve. Adequate head room should be allowed for the rising stem when installing valves of this type.

Figure 5.2a illustrates the bonnet design most commonly used for gate, globe, and angle valves of the smaller sizes.

Figure 5.2c illustrates a nonrising stem with an inside screw. This type of stem can be used where head room is limited and where corrosive action of liquids or gases in the line is not a problem. Also, there is a minimum amount of wear on the stem packing.

The draftsman should have access to catalogs of the various valve manufacturing companies. If he is to become proficient in valve selection, he should become familiar with all the variations in valve design and know how to specify and draw these variations. His drawings and specifications should leave no possible room for misunderstanding.

END CONNECTIONS

In addition to the basic valve design, a knowledge of the various end connections on valves is important.

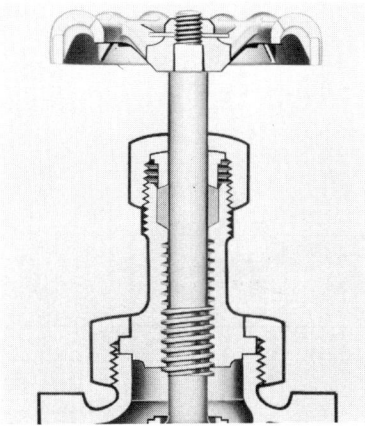

FIG. 5.2. STEM VARIATIONS

Fig. 5.2a. Rising stem with inside screw. (Courtesy Crane Co.)

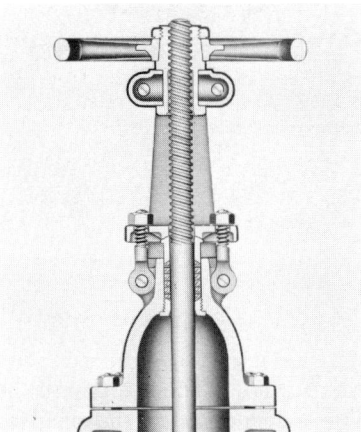

Fig. 5.2b. Stem rises, handwheel does not. (Courtesy Crane Co.)

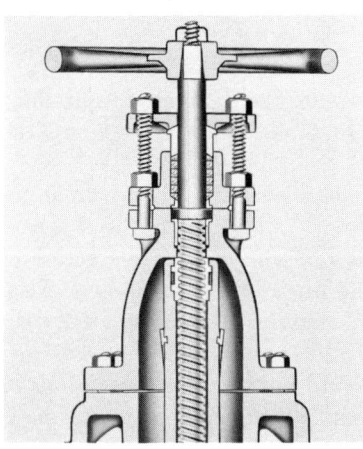

Fig. 5.2c. Nonrising stem with inside screw. (Courtesy Crane Co.)

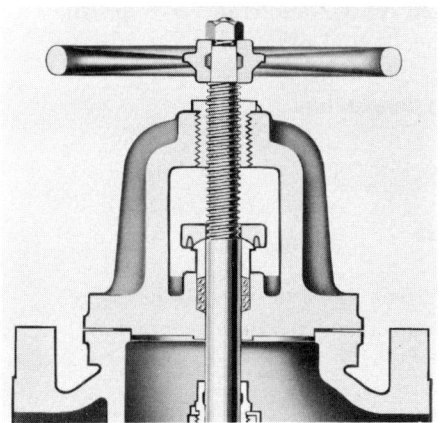

Fig. 5.2d. Handwheel rises with stem. (Courtesy Crane Co.)

FUNDAMENTALS OF PIPE DRAFTING

Figure 5.3 illustrates the common types of end connections used on valves.

Threaded ends. The threaded-end globe valve illustrated in Fig. 5.3a represents a common type of valve with screwed ends. (It is common practice to refer to threaded valves and fittings as "screwed.") Threaded connections are found in brass, iron, steel, and in many alloy piping materials. They may be used for all ordinary pressures, but their use is usually limited to the smaller pipe sizes.* Larger size pipe is difficult to make up with screwed joints.

Solder ends. The solder end joint, one of which is illustrated on the globe valve in Fig. 5.3b, is used with copper tubing for plumbing and heating lines and for many low-pressure industrial services.

Welding ends. In this discussion, the student should keep in mind that the reference is to valves only, and not to fittings in general.

Two types of welding ends on valves are available, butt-welding and socket-welding ends, the socket-welding ends usually being limited to the smaller sizes. Welding-end valves are obtainable only in steel and are used for higher pressure-temperature services, and on lines not requiring frequent dismantling. Figure 5.3c illustrates a valve designed to be attached to the line by means of a butt weld.

Flanged ends. Flanged-end valves, one of which is illustrated in Fig. 5.3d, although made in sizes as small as ½ inch, are generally used for larger lines because they are easy to install or to remove from a line. The bodies of flanged valves are obtainable in various grades and alloys of steel or iron; however, the trim (stem, seat, inner valve, etc.) may usually be obtained in brass, bronze, or other alloys.

In selecting flanged-end valves, as in all valves, it is necessary to know the conditions under which the valve is to operate. Pressure, strain, corrosion, temperature encountered, and the desired function of the valve must all be considered. The flanged ends must also be selected so as to be in accord with the connecting flanges used in the system in regard to strength specifications, bolting, and material requirements.

Hub ends. Hub ends similar to the ones illustrated in Fig. 5.3e are generally limited to valves for water supply and sewage piping.

---

* Many pressures used in modern technology are far above what would be considered "ordinary." The use of threaded connections should therefore be controlled and limited to pressures that would be classified by the engineer as ordinary.

CONTROLS

VALVE MATERIALS

Another important phase of valve selection is selection of materials. Valves used in industry fall into three basic material groups: (1) brass, (2) iron, (3) steel. Each of these groups has one or more variants, each with its own particular service characteristic.

Brass. Brass, an alloy of copper and tin, should not be used for temperatures exceeding 450°F.

Bronze. Bronze, an alloy of copper and zinc, is somewhat stronger and tougher than brass. It is recommended for temperatures up to 550°F. Bronze valves are normally made in sizes up to 3 inches and for pressures up to 330 pounds per square inch. The use of bronze is recommended when salt solutions are being handled if the pressure and temperature involved will permit.

Note. Although the above discussion regarding the composition of brass and bronze is essentially correct, the terms have been so loosely used in the trade that their use has little descriptive value. In the selection of brass or bronze valves the recommendations of the manufacturer should be closely studied and followed.

Iron. Iron is made in several grades, the composition varying with the specifications of the valve manufacturing company. Its uses vary from cast iron used for small valves and fittings having light metal sections to high-strength metal alloy cast iron used for castings for large valves. Cast iron should not be used for temperatures exceeding 450°F.

Malleable iron also falls in the iron classification. It is particularly suited for screwed fittings and unions; however, it is used to some extent when tightness, stiffness, and toughness are desired. It is especially valuable for piping and parts subjected to expansion and contraction stresses and shock.

Steel. Steel is recommended for high pressures and temperatures and for conditions where stresses, either internal or external, may be too severe for iron

FIG. 5.3. END CONNECTIONS

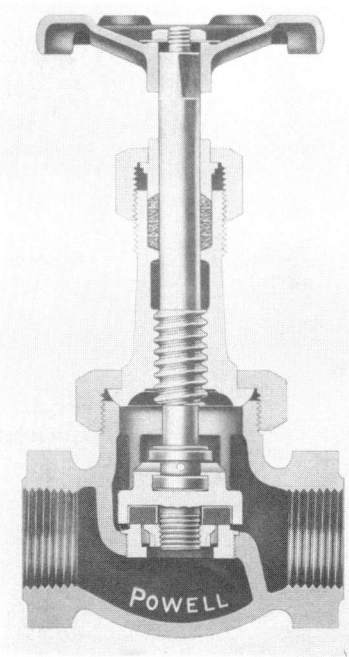

Fig. 5.3a. Globe valve, screwed end. (Courtesy Wm. Powell Co.)

Fig. 5.3b. Globe valve, solder joint. (Courtesy Wm. Powell Co.)

Fig. 5.3c. Lift-check valve, welding ends. (Courtesy Wm. Powell Co.)

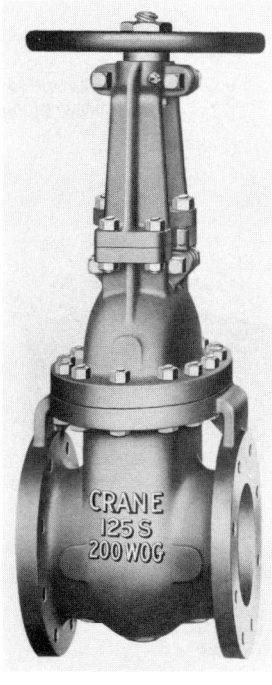

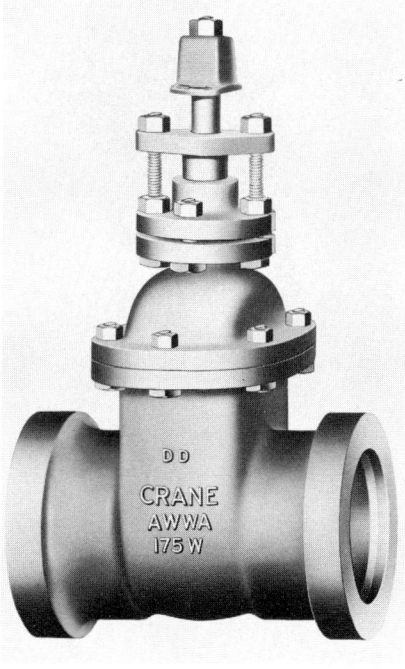

Fig. 5.3d. Gate valve, flanged ends. (Courtesy Crane Co.)

Fig. 5.3e. Gate valve, hub ends. (Courtesy Crane Co.)

FIG. 5.4. EQUIPMENT OFTEN

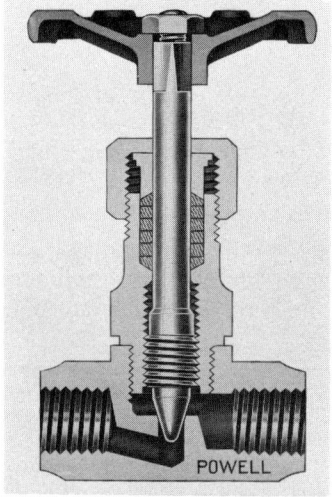

Fig. 5.4a. Needle-point valve. (Courtesy Wm. Powell Co.)

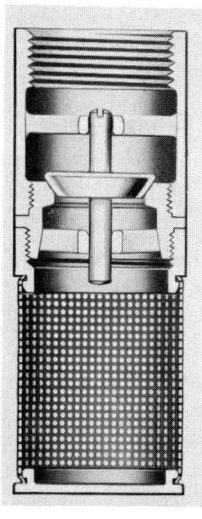

Fig. 5.4b. Foot valve with strainer. (Courtesy Crane Co.)

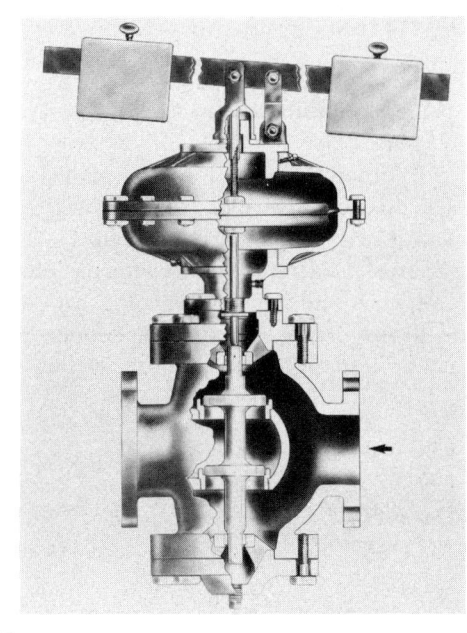

Fig. 5.4c. Back pressure or vacuum regulator. (Courtesy Black Sivalls & Bryson, Inc.)

Fig. 5.4d. Relief and safety valve. (Courtesy Wm. Powell Co.) ▼

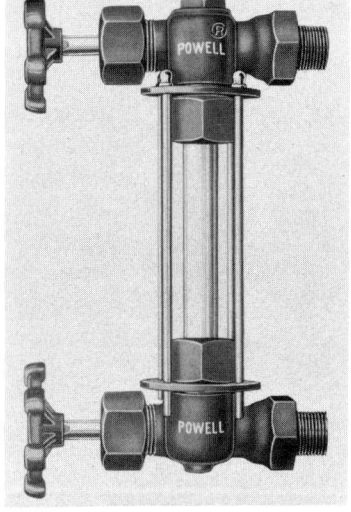

Fig. 5.4e. Liquid-level gage. (Courtesy Wm. Powell Co.)

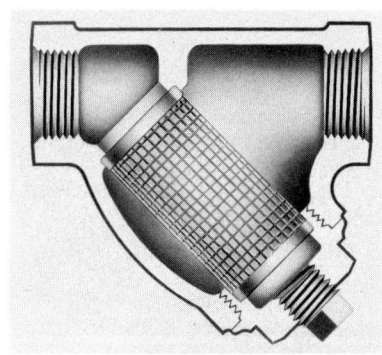

Fig. 5.4f. Sediment separator. (Courtesy Crane Co.)

Fig. 5.4g. Inverted open-float steam trap. (Courtesy Crane Co.) ◄

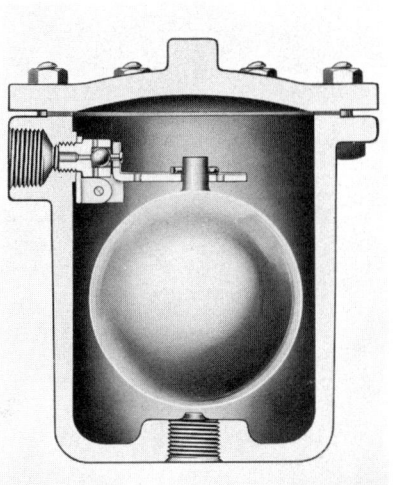

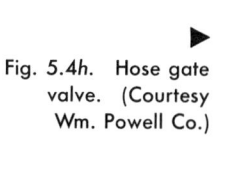

► Fig. 5.4h. Hose gate valve. (Courtesy Wm. Powell Co.)

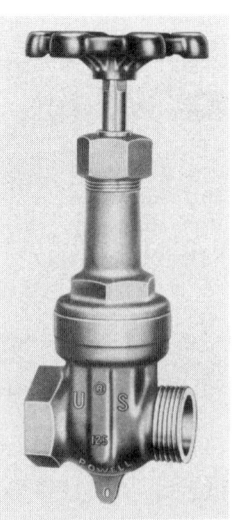

USED IN PIPING SYSTEMS

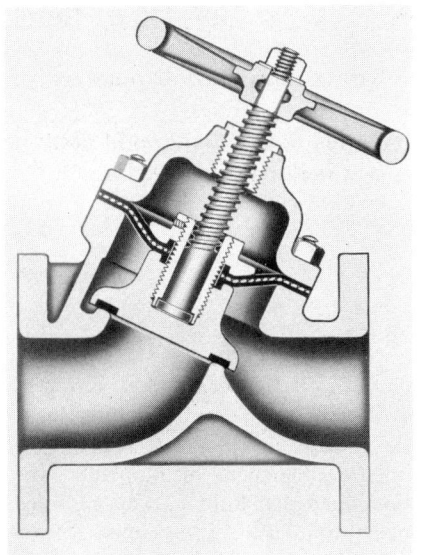

Fig. 5.4i. Packless diaphragm valve. (Courtesy Crane Co.)

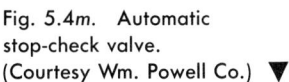

Fig. 5.4m. Automatic stop-check valve. (Courtesy Wm. Powell Co.) ▼

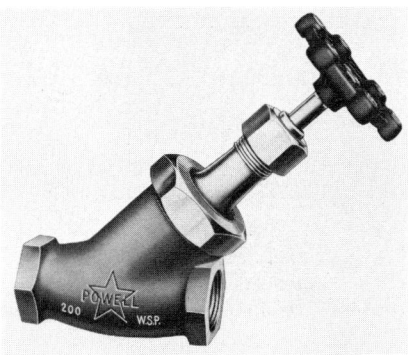

Fig. 5.4n. Diaphragm control valve. (Courtesy Black Sivalls & Bryson, Inc.) ▼

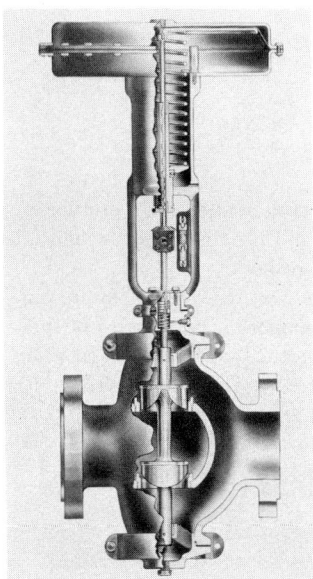

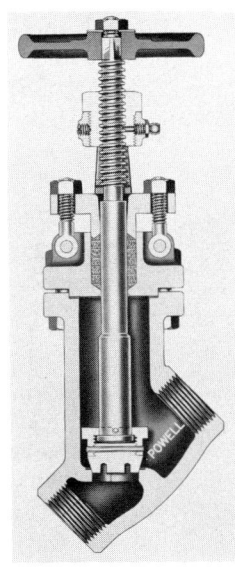

Fig. 5.4j. "Y" valve. (Courtesy Wm. Powell Co.)

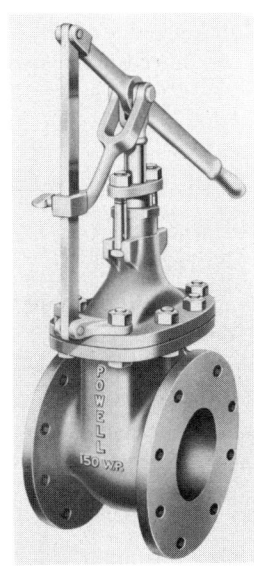

Fig. 5.4k. Quick-opening gate valve. (Courtesy Wm. Powell Co.)

Fig. 5.4l. Motor-operated valve. (Courtesy Wm. Powell Co.)

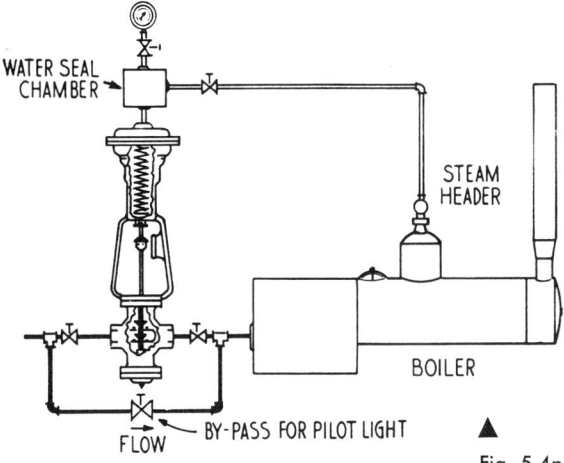

Fig. 5.4p. Boiler fuel governor.

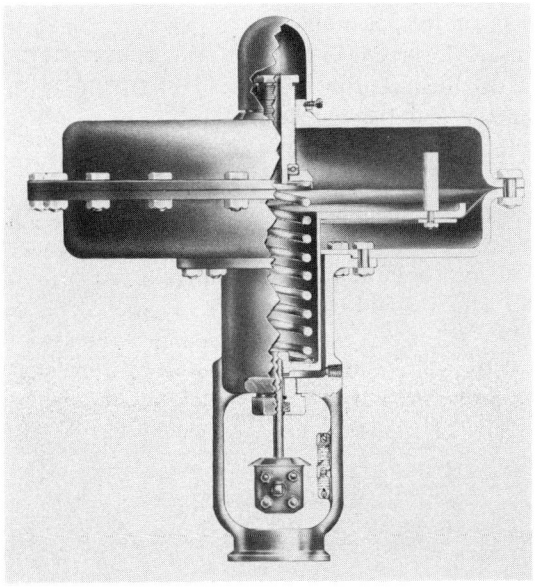

◄ Fig. 5.4o. Reverse top works, diaphragm control valve. (Courtesy Black Sivalls & Bryson, Inc.)

or brass. Steel may be cast or forged, and may be alloyed with strengthening or corrosion-resistant materials.

SPECIAL CONTROLS

Few industrial piping systems operate without the use of special controls and devices typical of the process involved. Figure 5.4 illustrates a few of the common valve variations. If no drawing symbol exists for the control in question, the usual practice is to make an outline drawing resembling the piece and to accompany the drawing with a note or description.

The method of valve control is an important consideration in the design of a piping system. The safety valve (Fig. 5.4d) operates automatically when pressure reaches a predetermined point. A float valve is actuated by a float which controls the valve; it is used to hold fluid at a constant level. Pressure regulators, of which Fig. 5.4c illustrates one of many designs, are usually spring- or weight-loaded diaphragm valves. Other valves are controlled by motors, as illustrated in Fig. 5.4l.

Another type of control valve is the diaphragm control valve. Pressure (usually air), is applied to a diaphragm, which in turn actuates the valve. The valve may be so designed as to be closed or opened by air pressure as the need may be.

Figure 5.4n illustrates a common type of diaphragm control valve. It may be designed so that its normal position is either open or closed. The open position is specified by the letter "D" following the type number, or if the valve is normally to remain closed, the letter "R" follows the type number. Also, the top works on this type of valve may be reversed causing the spring mechanism to appear above instead of below the diaphragm, as illustrated in Fig. 5.4n. If a reverse top works, as illustrated in Fig. 5.4o, is desired, the type number and the normal position letter are followed by the letters RT, the abbreviation for reversed top works. The valve illustrated in Fig. 5.4n is designated by the manufacturer as Type 86D.

The top works illustrated in Fig. 5.4o could replace the top work of Fig. 5.4n which would change the description to Type 86DRT.

It should be pointed out that the valve illustrated is designed so as to utilize two ports, the purpose of which is to allow for a free flow of liquid or gas through the valve with a minimum of inner valve movement.

Figure 5.4p illustrates how a diaphragm control valve may be used as a boiler feed governor.

Problem

Problem 1. Figure 5.5 is the plan and elevation of a room which is to contain the pumping station for a food-processing plant. Two pumps are to serve three tanks which are located in other parts of the plant. The piping is to be so arranged that fluid may be pumped from tank to tank or be drained from the tanks. Steam will be used as a cleaning agent. All lines leading from the discharge sides of the pumps are to be 3″, and all lines leading to the suction sides of the pumps are to be 4″. All steam lines leading to the main lines (both discharge and suction) are to be 2″. Use cast-iron OS&Y flanged gate valves on all 3″ and 4″ lines and screwed globe valves on all steam lines.

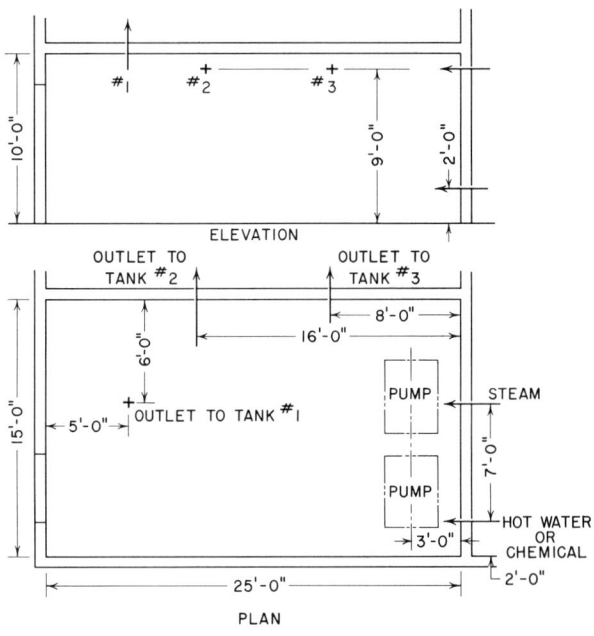

Fig. 5.5.

Figure 5.6 is a hookup recommended for a similar situation. Study this hookup and the discussion which follows it and adapt it to this problem.

Figure 5.7 is an outline drawing of the motor and pump to be used. The dimensions given are not those of any particular pump or motor but may be used to obtain an idea as to the space occupied by the pumps and motors.

CONTROLS

31

Lay out the pipe in single-line orthographic indicating all pipe sizes, types of valves, relative location of orifice plates, and other information necessary for the solution of the problem. Do not indicate center-to-center distances, but show the piping in approximately its correct positions in relation to walls, ceiling, and floors. Consider the location of valves with respect to accessibility and convenience for operation. Represent all valves with the correct symbol and add additional information as to size and type by note.

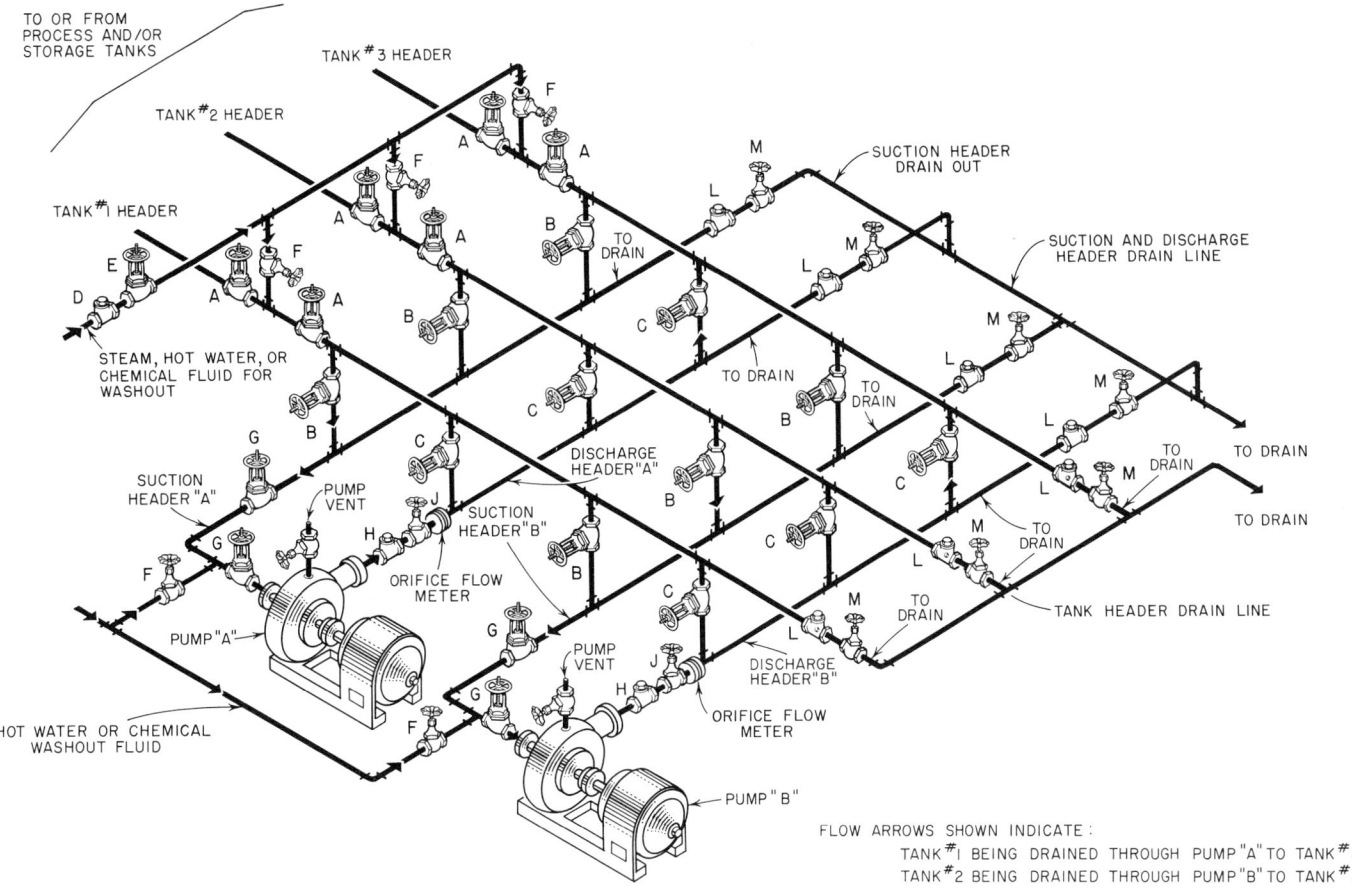

Valve Recommendations

Code	Quan.	Type	Application
A	6	Gate	Tank header for cleaning
B	6	Gate	Suction header inlet
C	6	Gate	Discharge header outlet
D	1	Swing Check	Prevent back flow to washout line
E	1	Gate	Washout line shutoff
F	5	Globe	Individual washout shutoffs to tank headers
G	4	Gate	Pump suction shutoffs
H	2	Swing Check	Prevent back flow to pump
J	2	Globe	Pump discharge control
K	2	Globe	Pump vent control
L	7	Swing Check	

Fig. 5.6. Central pumping station for transfer of fluids. Transfer of fluids within a plant is a major operation in many industries such as food, distilling, brewing, pasteurizing, and chemical. The flexible pumping station lay-out shown here, which can be applied to any of these industries and other similar ones, allows fluids to be pumped from tank to tank or to any processing equipment. It also reduces piping to a minimum and eliminates the nuisance of continual hose coupling.

A single pipeline from each tank or piece of processing equipment is run to the pumping station and there connected to the suction and discharge headers of each pump. Although the number of pumps and tank headers may vary according to the plant needs, each tank header must be tied into the suction and discharge headers of all pumps.

Because pollution of one fluid by the remains of the previously pumped fluid is often harmful to the product, adequate provision must be made for washing pumps, valves, and pipelines. This may be done by steam, hot water, or chemical solutions, depending upon the fluids carried. All headers must have proper drainage facilities and check valves should be installed on drain lines to prevent destructive backflow into the headers. Flow meters are often installed in the pump discharges to measure or regulate the rate of flow and quantity of fluid. (Courtesy Jenkins Bros.)

32 FUNDAMENTALS OF PIPE DRAFTING

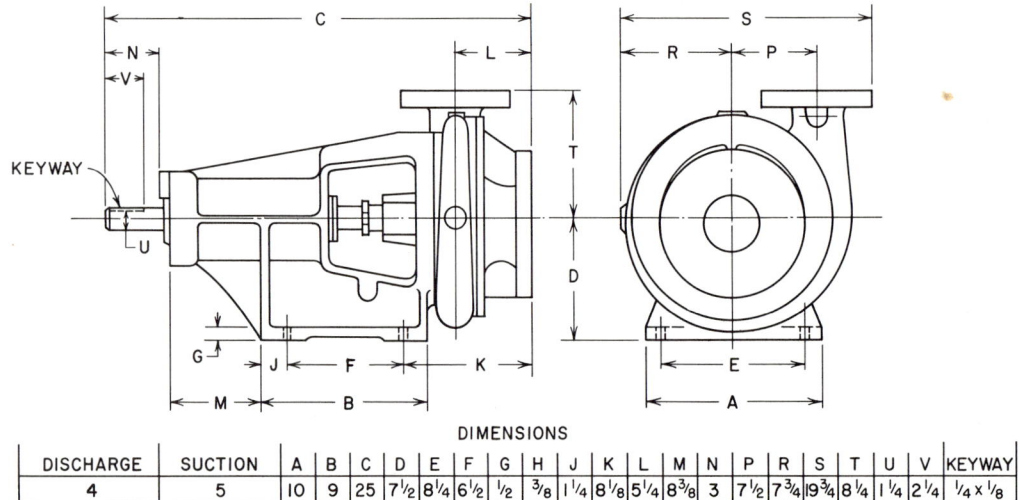

DIMENSIONS

DISCHARGE	SUCTION	A	B	C	D	E	F	G	H	J	K	L	M	N	P	R	S	T	U	V	KEYWAY
4	5	10	9	25	7 1/2	8 1/4	6 1/2	1/2	3/8	1 1/4	8 1/8	5 1/4	8 3/8	3	7 1/2	7 3/4	19 3/4	8 1/4	1 1/4	2 1/4	1/4 x 1/8

Fig. 5.7a. Centrifugal pump.

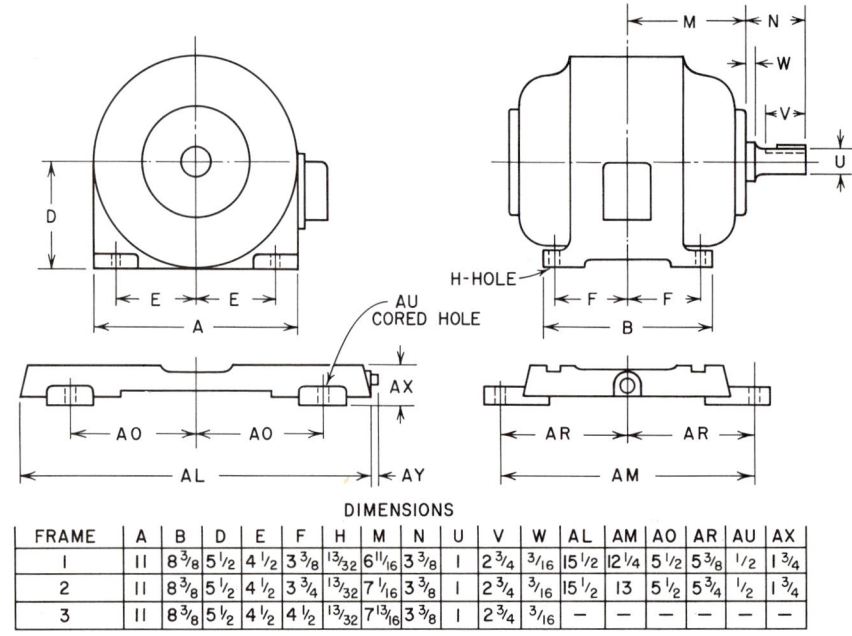

DIMENSIONS

FRAME	A	B	D	E	F	H	M	N	U	V	W	AL	AM	AO	AR	AU	AX
1	11	8 3/8	5 1/2	4 1/2	3 3/8	13/32	6 11/16	3 3/8	1	2 3/4	3/16	15 1/2	12 1/4	5 1/2	5 3/8	1/2	1 3/4
2	11	8 3/8	5 1/2	4 1/2	3 3/4	13/32	7 1/16	3 3/8	1	2 3/4	3/16	15 1/2	13	5 1/2	5 3/4	1/2	1 3/4
3	11	8 3/8	5 1/2	4 1/2	4 1/2	13/32	7 13/16	3 3/8	1	2 3/4	3/16	—	—	—	—	—	—

Fig. 5.7b. Motor.

6 Pipe and pipe fittings

PIPE

The selection of materials to be used for the pipe of a system is important because piping systems are subjected to varying conditions of heat, strain, pressure, and chemical reactions. Cast iron, steel, and wrought iron are the most common pipe materials.

Cast iron is suitable for underground installations such as gas and water mains, for plumbing lines, or for low-pressure steam systems. Cast iron has a high resistance to corrosion, but has small resistance to strain and shock.

Steel pipe is resistant to high pressure and temperature, especially when other metals such as nickel or chromium are added. Steel pipe is also adaptable to various types of fittings and connections.

Wrought iron is costlier than steel and not as strong, but when corrosive conditions are present, it is considered by some designers to be preferable to steel.

For handling liquids containing salts, seamless brass pipe or tubing is sometimes substituted for wrought iron or steel pipe; however, the cost is considerable, and for this reason, brass should be used only when conditions justify the extra expense.

Copper pipe or tubing will also withstand corrosion, but will not withstand conditions of high pressure and repeated stress.

Lead pipe and lead-lined pipe is sometimes used if it is to be subjected to the action of acids.

Ordinary steel pipe which has been dipped in molten zinc to prevent rust is called galvanized pipe and is suitable for lines containing drinking water.

Plastic pipe may be used when flexibility and ease of installation are desired. Plastic pipe is resistant to acid and chemical substances, but should not be used under extreme heat or pressure. Ordinarily, plastic pipe does not affect the taste of water. Unless specifically recommended by the manufacturer, it should not be brought in contact with oil. This pipe is very satisfactory for underground installation.

Seamless flexible metal tubing is used on equipment where vibration is present. When tubing is made in thicknesses corresponding to standard steel pipe it is called pipe.

PIPE SIZING

Wrought-iron and steel pipe are specified according to the nominal inside diameter in all sizes under 14 inches, and is manufactured in various weights to meet different strength requirements. The most common weights are: standard, extra strong, and double extra strong; however, there are many other weights known by trade names such as hydraulic pipe, merchant casing, API (American Petroleum Institute) pipe, etc.

A series of schedule numbers are used by the American Standards Association as a means of specifying wall thicknesses. These schedule numbers are an approximation of the values obtained by applying the formula $1,000 \times (P/S)$ = Schedule Number, where P is gage pressure (internal pressure), and S is the allowable fibre stress in pounds per square inch (psi). Recommended values of S may be found in tables published by pipe manufacturers.

S values for seamless and electric resistance welded

carbon-steel pipe of several material specifications and temperature conditions can be found in the table in Fig. 6.1. This is one of many tables designed for the purpose of indicating S values of the many materials and conditions under which pipe may be used.

When the schedule number is obtained by applying the above formula, all measurements of the pipe can be found by reference to Fig. 6.2. There are, of course, other factors involved, a discussion of which is beyond the scope of this book. The draftsman should follow closely the recommendations of the engineer.

All pipe under 14 inches is designated by stating the nominal inside diameter and schedule number. All pipe over 14 inches is designated by stating the actual outside diameter and thickness of the walls.

Tubing of all sizes is specified by stating the outside diameter, wall thickness, and weight per foot.

MATERIAL DESCRIPTION		CARBON STEEL[2]								
		SEAMLESS				ELECTRIC RESISTANCE WELDED				
ASME BOILER CODE		x	x	x	–	x	x	x	x	x
ASA POWER PIPING		x	x	x	x	x	–	x	–	–
ASA DISTRICT HEATING[10]		x	x	x	x	x	–	x	–	–
COMMON IDENTIFICATION		Grade A		Grade B	Ordinary	Grade A		Grade B	O.H. Iron	
ASTM MATERIAL SPECIFICATION NUMBER		A53 Gr A A106 Gr A	A83 Type A[4] A179 LowC[13] A192	A53 Gr B A106 Gr B A210	A120	A53 Gr A A135 Gr A	A178 Gr A[4] A226	A53 Gr B A135 Gr B	A178 Gr B	A178 Gr C
MINIMUM ULTIMATE STRENGTH (psi)		48,000	47,000	60,000		48,000	47,000	60,000	40,000	60,000
NOTES					(7)	(6)		(7)		(4) (6)
METAL TEMPERATURE °F[1]	–20/100 200 300 400[9]	12,000	11,750	15,000	10,800 10,600 10,700 9,800	10,200	11,750	12,750	8,500	15,000
	450 500 600 650 700	↓ 12,000 11,650	↓ 11,750 11,500[15]	↓ 15,000 14,350	9,600	↓ 10,200 9,900	↓ 11,750 11,500	↓ 12,750 12,200	↓ 8,500 8,200	↓ 15,000 14,350
	750 800 850 900	10,700 9,000 7,100 5,000	10,700[15] 9,000 7,100 5,000	12,950 10,800 7,800 5,000		9,100 9,650[12] 6,050[12] 4,250[12]	9,100 7,650 6,050 4,250	11,000 9,200[12] 6,650[12] 4,250[12]	7,400	11,000 9,200 6,650 4,250
	950 1000	3,000[11] 1,500[11]	3,000[11] 1,500[11]	3,000[11] 1,500[11]			2,550 1,300			2,550 1,300

x in block indicates that the material is listed in the code section.

– (dash) in block indicates that the material is *not* listed in the code section.

[1] Allowable S values for intermediate temperatures may be obtained by interpolation.

[2] Upon prolonged exposure to temperatures above about 800°F, the carbide phase of carbon steel may be converted to graphite.

[3] Upon prolonged exposure to temperatures above 875°F, the carbide phase of carbon-molybdenum steel may be converted to graphite.

[4] Only killed steel shall be used above 900°F.

[5] At temperatures from 200°F through 1050°F these stresses meet all of the criteria specified for establishing stress values, except that they exceed 62½ per cent but do not exceed 90 per cent of the yield strength at temperature. They may be used where slightly greater deformation is permissible.

[6] Above 700°F these stress values include a joint efficiency factor of .0.85. When material to this specification is used for pipe, multiply the stress values up to and including 700°F by a factor of 0.85.

[7] For electric-resistance-welded pipe for applications, where the temperature is below 650°F, and where pipe furnished under this classification is subjected to supplemental tests and/or heat treatment as agreed by the supplier and purchaser, and whereby such supplemental tests and/or heat treatments demonstrate the strength characteristics of the weld to be equal to the minimum tensile strength specified for the pipe, the S values equal to the corresponding seamless grades may be used.

[8] Cast-iron pipe shall not be used for lubricating oil lines for machinery and in any case not for oil having a temperature above 300°F.

[9] For Power Piping and District Heating, the S values given at 400°F may be used for steam at 250 psi (406°F).

[10] District Heating stresses apply only at temperatures of 750°F and below.

[11] These stresses are for A106, A192 and A210 in ASME Boiler Code only.

[12] These stresses are for ASME Boiler Code only.

[13] Not in ASME Boiler Code.

[14] These stresses apply only up to 500°F for District Heating piping.

[15] For ASA Power and District Heating the allowable stress for 700°F is 11,450 psi and for 750°F is 10,550 psi.

Fig. 6.1. ASME power boiler, ASA power piping and district heating systems. Allowable S values (psi). (Reproduced by courtesy of Tube Turns, Louisville, Kentucky, copyright 1955.)

PIPE AND PIPE FITTINGS

NOMINAL PIPE SIZE	OUT-SIDE DIAM.	SCHED. 5*	SCHED. 10*	SCHED. 20	SCHED. 30	STAND-ARD†	SCHED. 40	SCHED. 60	EXTRA STRONG§	SCHED. 80	SCHED. 100	SCHED. 120	SCHED. 140	SCHED. 160	XX STRONG
⅛	0.405	.....	0.049	.....	.....	0.068	0.068	.....	0.095	0.095	.....	.....	.....	.....	.....
¼	0.540	.....	0.065	.....	.....	0.088	0.088	.....	0.119	0.119	.....	.....	.....	.....	.....
⅜	0.675	.....	0.065	.....	.....	0.091	0.091	.....	0.126	0.126	.....	.....	.....	.....	.....
½	0.840	.....	0.083	.....	.....	0.109	0.109	.....	0.147	0.147	.....	.....	.....	0.187	0.294
¾	1.050	0.065	0.083	.....	.....	0.113	0.113	.....	0.154	0.154	.....	.....	.....	0.218	0.308
1	1.315	0.065	0.109	.....	.....	0.133	0.133	.....	0.179	0.179	.....	.....	.....	0.250	0.358
1¼	1.660	0.065	0.109	.....	.....	0.140	0.140	.....	0.191	0.191	.....	.....	.....	0.250	0.382
1½	1.900	0.065	0.109	.....	.....	0.145	0.145	.....	0.200	0.200	.....	.....	.....	0.281	0.400
2	2.375	0.065	0.109	.....	.....	0.154	0.154	.....	0.218	0.218	.....	.....	.....	0.343	0.436
2½	2.875	0.083	0.120	.....	.....	0.203	0.203	.....	0.276	0.276	.....	.....	.....	0.375	0.552
3	3.5	0.083	0.120	.....	.....	0.216	0.216	.....	0.300	0.300	.....	.....	.....	0.438	0.600
3½	4.0	0.083	0.120	.....	.....	0.226	0.226	.....	0.318	0.318	.....	.....	.....	.....	.....
4	4.5	0.083	0.120	.....	.....	0.237	0.237	.....	0.337	0.337	.....	0.438	.....	0.531	0.674
5	5.563	0.109	0.134	.....	.....	0.258	0.258	.....	0.375	0.375	.....	0.500	.....	0.625	0.750
6	6.625	0.109	0.134	.....	.....	0.280	0.280	.....	0.432	0.432	.....	0.562	.....	0.718	0.864
8	8.625	0.109	0.148	0.250	0.277	0.322	0.322	0.406	0.500	0.500	0.593	0.718	0.812	0.906	0.875
10	10.75	0.134	0.165	0.250	0.307	0.365	0.365	0.500	0.500	0.593	0.718	0.843	1.000	1.125	.....
12	12.75	0.156	0.180	0.250	0.330	0.375	0.406	0.562	0.500	0.687	0.843	1.000	1.125	1.312	.....
14 O.D.	14.0	.....	0.250	0.312	0.375	0.375	0.438	0.593	0.500	0.750	0.937	1.093	1.250	1.406	.....
16 O.D.	16.0	.....	0.250	0.312	0.375	0.375	0.500	0.656	0.500	0.843	1.031	1.218	1.438	1.593	.....
18 O.D.	18.0	.....	0.250	0.312	0.438	0.375	0.562	0.750	0.500	0.937	1.156	1.375	1.562	1.781	.....
20 O.D.	20.0	.....	0.250	0.375	0.500	0.375	0.593	0.812	0.500	1.031	1.281	1.500	1.750	1.968	.....
22 O.D.	22.0	.....	0.250	.....	.....	0.375	.....	.....	0.500	.....	.....	.....	.....	.....	.....
24 O.D.	24.0	.....	0.250	0.375	0.562	0.375	0.687	0.968	0.500	1.218	1.531	1.812	2.062	2.343	.....
26 O.D.	26.0	.....	.....	.....	.....	0.375	.....	.....	0.500	.....	.....	.....	.....	.....	.....
30 O.D.	30.0	.....	0.312	0.500	0.625	0.375	.....	.....	0.500	.....	.....	.....	.....	.....	.....
34 O.D.	34.0	.....	.....	.....	.....	0.375	.....	.....	0.500	.....	.....	.....	.....	.....	.....
36 O.D.	36.0	.....	.....	.....	.....	0.375	.....	.....	0.500	.....	.....	.....	.....	.....	.....
42 O.D.	42.0	.....	.....	.....	.....	0.375	.....	.....	0.500	.....	.....	.....	.....	.....	.....

All dimensions are given in inches.

The decimal thicknesses listed for the respective pipe sizes represent their nominal or average wall dimensions. The actual thicknesses may be as much as 12.5 per cent under the nominal thickness because of mill tolerance. Thicknesses shown in light face for Schedule 60 and heavier pipe are not currently supplied by the mills, unless a certain minimum tonnage is ordered.

*Thicknesses shown in *italics* are for Schedules 5S and 10S, which are available in stainless steel only.

†Thicknesses shown in *italics* are available also in stainless steel, under the designation Schedule 40S.

§Thicknesses shown in *italics* are available also in stainless steel, under the designation Schedule 80S.

Fig. 6.2. Dimensions of seamless and welded steel pipe. This table lists the pipe sizes and wall thicknesses currently established as standard, or specifically: (1) The traditional standard weight, extra strong and double extra strong pipe. (2) The pipe wall thickness schedules listed in American Standard B36.10, which are applicable to carbon steel and alloys *other than* stainless steels. (3) The pipe wall thickness schedules listed in American Standard B36.19, which are applicable *only* to stainless steels. (Reproduced by courtesy of Tube Turns, Louisville, Kentucky, copyright 1952.)

PIPE FITTINGS

Pipe fittings are parts used in making up pipe. They may be screwed, welded, soldered, or flanged. There are also many patented fittings designed for special purposes or with the idea of offering certain advantages.

Screwed connections. Screwed fittings are manufactured in steel, cast iron, malleable iron, and bronze, each of which is available in several service ratings. The working pressures and general conditions existing in a line should be a guide to the draftsman in the selection of fittings. They should have the same resistive qualities as that of the pipe they join. Figure 2.10a is a typical catalog illustration showing measurements of 125 and 250 psi cast-iron fittings.

Generally speaking, pipe sizes of 2 inches and under use threaded fittings; however, some designers believe that, when all things are considered, welding, even on small sizes, is more economical than threaded fittings. The development of socket-type welding fittings illustrated in Fig. 6.3 has been a factor in the increased use of welding on small-size pipe.

Regardless of the increased use of welding, there will always be situations in which the use of threaded connections is necessary.

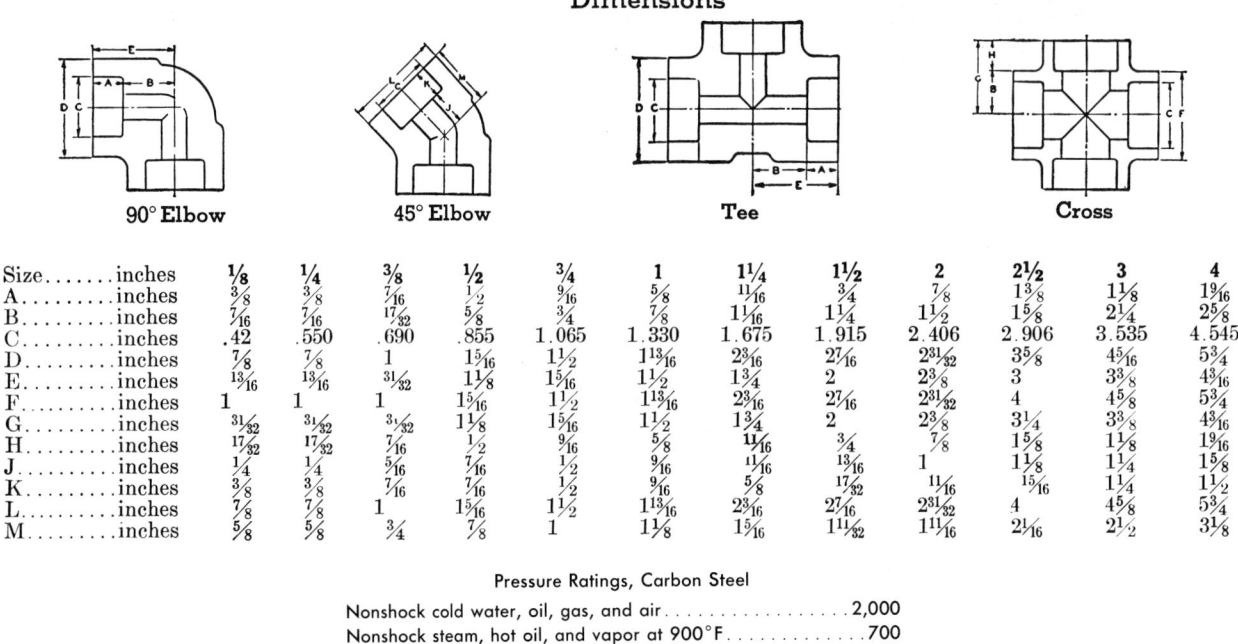

Dimensions

90° Elbow | 45° Elbow | Tee | Cross

Size......inches	1/8	1/4	3/8	1/2	3/4	1	1¼	1½	2	2½	3	4
A......inches	3/8	3/8	7/16	1/2	9/16	5/8	11/16	3/4	7/8	1 3/8	1 7/8	1 9/16
B......inches	7/16	7/16	17/32	5/8	3/4	7/8	1 1/16	1 1/4	1 1/2	1 5/8	2 1/4	2 5/8
C......inches	.42	.550	.690	.855	1.065	1.330	1.675	1.915	2.406	2.906	3.535	4.545
D......inches	7/8	7/8	1	1 5/16	1 1/2	1 13/16	2 3/16	2 7/16	2 31/32	3 5/8	4 5/16	5 3/4
E......inches	13/16	13/16	31/32	1 1/8	1 5/16	1 1/2	1 3/4	2	2 3/8	3	3 3/8	4 3/16
F......inches	1	1	1	1 5/16	1 1/2	1 13/16	2 3/16	2 7/16	2 31/32	4	4 5/8	5 3/4
G......inches	31/32	31/32	31/32	1 1/8	1 5/16	1 1/2	1 3/4	2	2 3/8	3 1/4	3 3/8	4 3/16
H......inches	17/32	17/32	7/16	1/2	9/16	5/8	11/16	3/4	7/8	1 5/8	1 1/8	1 9/16
J......inches	1/4	1/4	5/16	7/16	7/16	9/16	11/16	13/16	1	1 1/8	1 1/4	1 5/8
K......inches	3/8	3/8	7/16	7/16	1/2	9/16	5/8	17/32	11/16	15/16	1 1/4	1 1/2
L......inches	7/8	7/8	1	1 5/16	1 1/2	1 13/16	2 3/16	2 7/16	2 31/32	4	4 5/8	5 3/4
M......inches	5/8	5/8	3/4	7/8	1	1 1/8	1 5/16	1 11/32	1 11/16	2 1/16	2 1/2	3 1/8

Pressure Ratings, Carbon Steel

Nonshock cold water, oil, gas, and air.................2,000
Nonshock steam, hot oil, and vapor at 900°F............700

Fig. 6.3. Forged-steel socket welding fittings, 2,000 pound, for use with Schedule 40 (standard pipe). The bore of the fitting corresponds to the inside diameter of the designated schedule pipe. (Courtesy Walworth Co.)

Flanged connections. In pipe sizes larger than 2 inches it is common practice to use flanged connections on valves and fittings. A flange may be cast or forged as a part of the valve, fitting, or equipment on which it is used, but the mating flange which connects the pipe to the valve, fitting, or equipment may be attached to the pipe by welding as illustrated in Figs. 6.4a and 6.4b, by the use of a lap joint as illustrated in Fig. 6.4c, or by threads as illustrated in Fig. 6.4d.

The notes accompanying the illustrations of Fig. 6.4 explain briefly the normal uses and limitations of the common flange types.

Flanged fittings are normally made of cast iron or steel. Cast-iron fittings should not be used where they may be subjected to shock. In areas where there is danger of fire, cast iron should not be considered because a hot cast-iron part is apt to crack if struck by a water stream during a fire.

Cast-iron flanged fittings and flanges are made in three classes, 25 pound, 125 pound, and 250 pound. The 25 pound and 125 pound are identical in dimensions. The 250 pound have a 1/16-inch raised face.

Cast-steel fittings and forged-steel flanges are available in all ASA ratings from 150 pounds to 2,500 pounds.

Flange facings. Equal in importance with the type of flange used is the facing machined on the flange. Typical flange facings are illustrated in Fig. 6.5. The discussion pertaining to these facings is very general, but may be used as a guide when selecting flanges.

Welding fittings. Welding fittings may be used in place of either cast-iron or steel fittings. They are used extensively in prefabrication. Steam lines should be made up with welding fittings because of their strength and leak-resistant qualities. If a system is subjected to frequent dismantling, enough ordinary joints may be used to provide for ease of disassembly. Figure 2.10b illustrates welding fittings designed for use with standard pipe.

Problems

Problem 1. The distribution system illustrated in Fig. 6.6 is to be designed to utilize steam at a pressure of 530 psi at a temperature of 550°F. The steam engine is to operate at a pressure of 100 psi. The steam pressure for heating purposes is to be reduced to 5 psi. All high-pressure lines are 3″ pipe, all medium-pressure lines 2″. All low-pressure lines are to utilize welded fittings as much as possible with provisions for disassembly for replacements.

PIPE AND PIPE FITTINGS

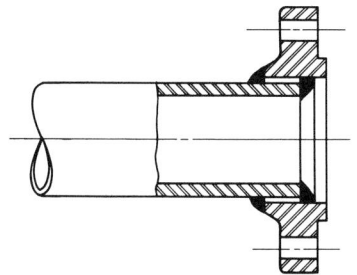 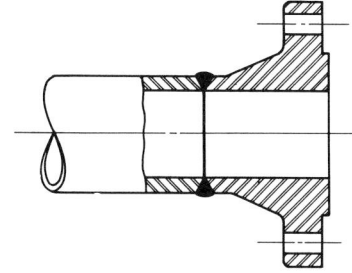 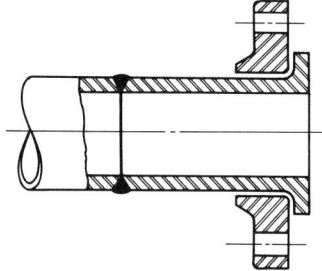

(a) Slip-on flange. Available in all pressure ratings, but most acceptable for the 150- and 300-pound classes. This flange is convenient when space does not permit the use of the butt-welding type.

(b) Welding-neck flange. Acceptable in all services through 2,500 pounds. This flange is recommended where pressure, temperature, shock, vibration, or external stress is high. In general, it has the most economical installed cost.

(c) Lap-joint flange. Recommended when alignment of bolts with a fixed flange is necessary. It is not recommended for large or thin-walled pipe.

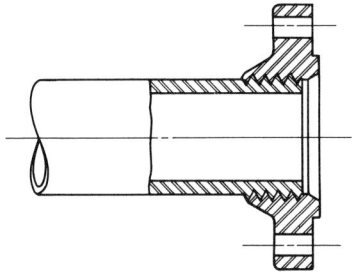 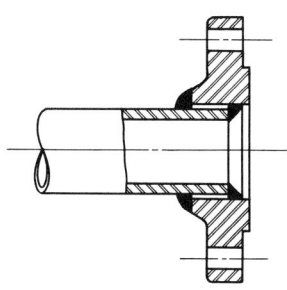 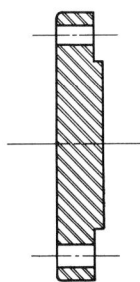

(d) Screwed flange. Most commonly used on the smaller pipe sizes, preferably 1½" and under. This flange should be seal welded for higher pressure services. It is useful when pipe and flange material are not suitable for welding.

(e) Reducing flange. Commonly used when change of line size is necessary in minimum space. Not recommended where it will be subjected to stress or critical flow characteristics.

(f) Blind flange. Used for line closure if future connection is anticipated.

Fig. 6.4. Service recommendations for flanges.

Using the formula for finding schedule numbers, $1,000 \times P/S$, and the "S" values found in Fig. 6.1 for carbon steel grade "A," specification ASTM A106, calculate the schedule number and make strength recommendations for all pipe. Prepare these calculations so they can be presented with the drawing.

Make a single-line isometric diagram representing all valves with their correct symbols. Indicate all pipe sizes and fittings by notes adjacent to the pipe or part. Include descriptive information on all valves as well as material recommendations.

Problem 2. The study of fluid flow requires considerable experience in the study of mathematics, hydraulics, and engineering in general, but the less complicated problems may often be left to the draftsman for solution; therefore, many common-sense principles of design should always be kept in mind. Friction of fluid or gas against the walls of pipe causes a certain calculable restriction of flow. It is sufficient at the present time to understand that fluid flow through a pipe cannot be calculated accurately unless friction is taken into consideration. It should be pointed out that the longer the pipe the more restriction due to friction. It should be pointed out also, that every change of direction of the pipe causes increased restriction to flow by increasing friction and causing turbulence to develop within the pipe. Valves of different design cause restriction of flow in variable amounts depending upon the design of the valve. A gate valve offers less resistance to flow than does a globe valve.

Sharp turns in a line restrict flow more than do long-radius bends. When efficiency of operation is important, it follows that long-radius bends and fittings should be used in preference to the conventional sharp-turn screwed fittings.

If friction and turbulence are disregarded, it is apparent that fluid will pass through pipe in a volume proportional to the pressure and the sectional area of the pipe. If pressure on the fluid in two pipes is equal, it follows that the larger pipe will pass the larger volume of fluid. A pipe which has a cross-sectional area of 2 sq. in. will

FUNDAMENTALS OF PIPE DRAFTING

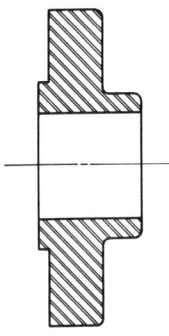

(a) Raised-face flange. Applicable to pressures to 900 lb.

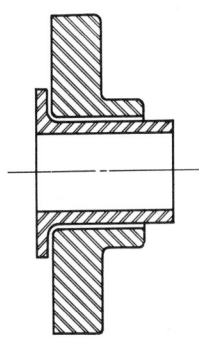

(b) Lap-joint flange. Any standard facing may be machined into upset face.

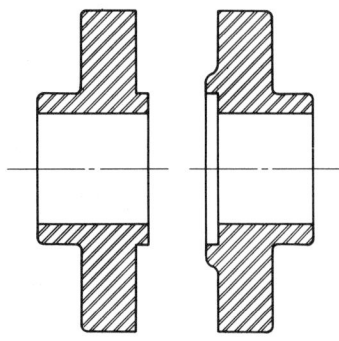

(c) Male and female flange (large). Use when desirable to have a facing that retains the gasket.

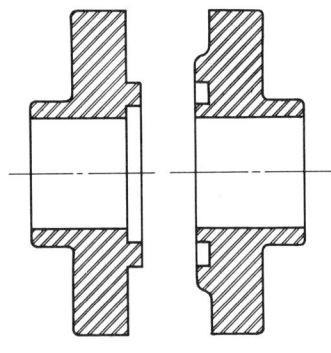

(d) Tongue and groove flange (large). Use when desirable to have a facing that retains the gasket.

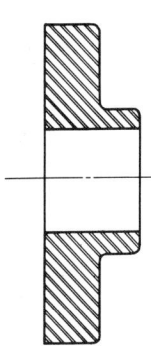

(e) Flat-face flange. Primarily used on cast iron.

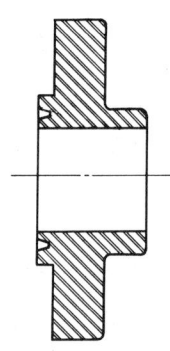

(f) Ring-joint flange. Use for high-pressure applications.

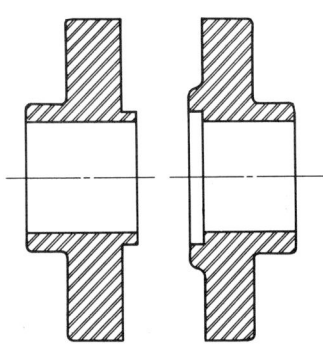

(g) Male and female flange (small). Same use as large male and female flange, except it provides for higher gasket compression.

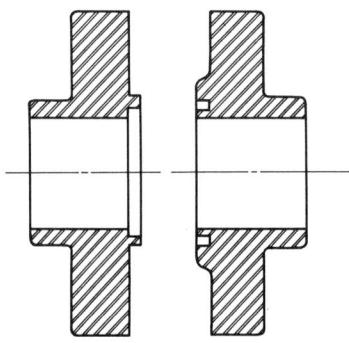

(h) Tongue and groove flange (small). Same use as large tongue and groove flange, except it provides for higher gasket compression.

Fig. 6.5. Typical flange facings.

PIPE AND PIPE FITTINGS

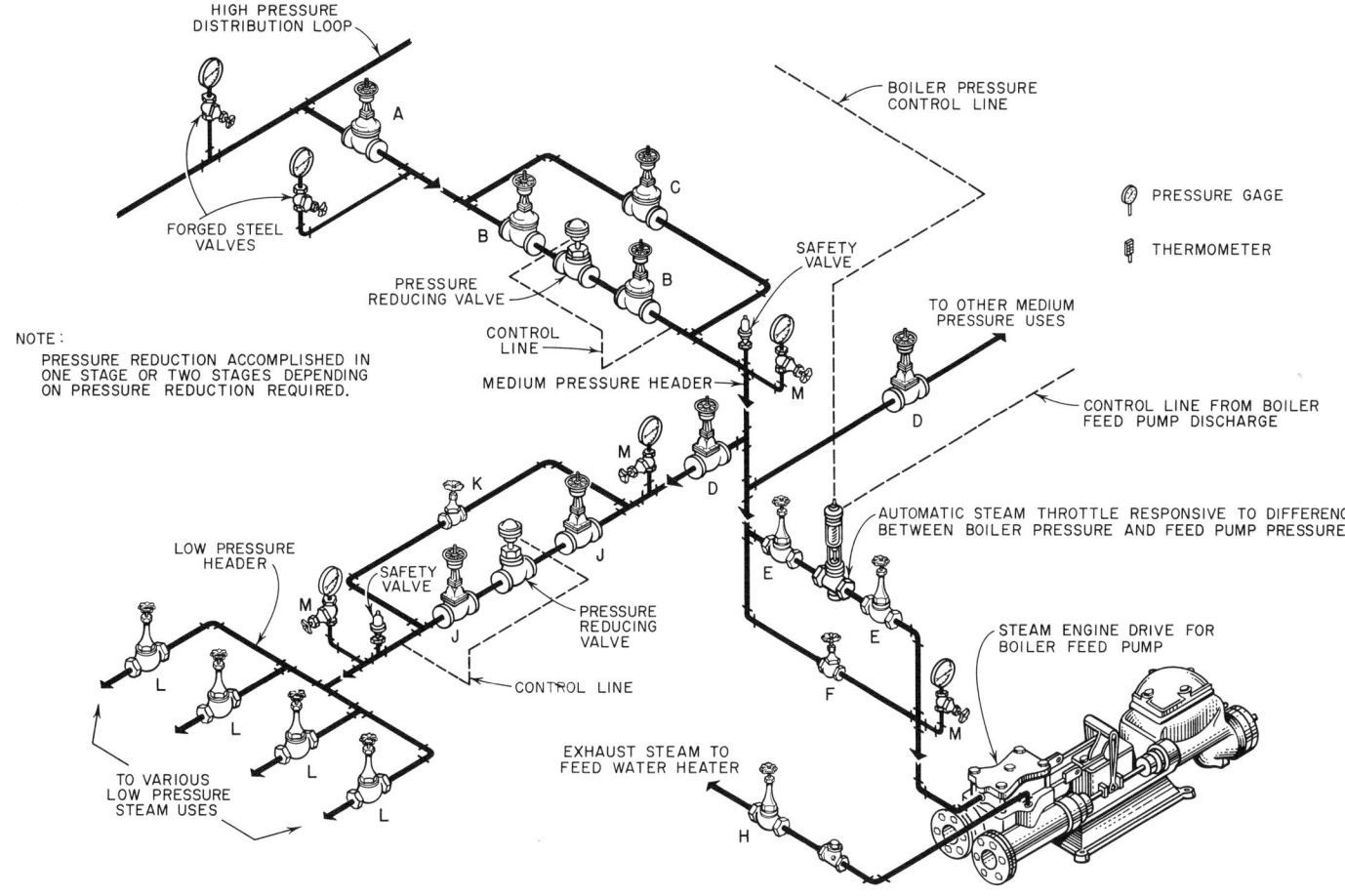

Valve Recommendations

Code	Quan.	Service
A	1	Shutdown of auxiliaries
B	2	Pressure-reducing valve shutoff
C	1	Manual control by-pass
D	2	Individual system shutdown
E	2	Automatic throttle shutoff
F	1	Manual control by-pass
G	1	Back-flow check
H	1	Steam-engine exhaust
J	2	Pressure-reducing valve shutoff
K	1	Manual control by-pass
L	4	Individual shutdowns
M	4	Pressure-gage control

Fig. 6.6. Medium- and low-pressure distribution from a high-pressure steam plant. When steam is generated at high pressure, provision must usually be made for medium- and low-pressure steam distribution. For example, steam generated at 450 psi might be reduced to 100 psi for the operation of steam-driven auxiliaries, and further reduced to 5 psi for heating purposes. The diagram shows a typical pressure-reduction hookup.

The two pressure-reducing stations are identical in lay-out. Each is provided with a pressure-reducing valve good for dead-end service, gate valves for shutoffs, and globe valves in the by-passes. Pressure gages are installed on both the high and the low side, and a safety valve on the low side is a guarantee against the possibility of pressure-reducing valve failure. Where large reductions in pressures are required, two stages are often advised, with reducing stations duplicated completely.

The boiler feed-pump steam-engine drive is controlled responsive to boiler feed-line pressure, and a manual by-pass around the control valve is provided.

Cast-steel gate and globe valves are recommended throughout the high-pressure system, and bronze and iron-body valves in the medium- and low-pressure systems. Outside screw and yoke gate valves with rising spindle are indicated where positive shutoff at infrequent intervals is required, and where it is necessary to determine quickly if the valve is open or closed. (Courtesy Jenkins Bros.)

pass twice as much fluid as a pipe which has a cross-sectional area of 1 sq. in.

In the following problem, it is necessary only to find the cross-sectional area of the standard-weight steel pipe, which has a nominal diameter of 4″. Pipes leading to the smaller tanks should thus be selected so that their cross-sectional areas are in proportion to the 600 cu. ft., 2,400 cu. ft., and 3,000 cu. ft. volumes of the tanks, and so that the sum of these areas is approximately equal to the cross-sectional area of the 4″ pipe.

Since standard-weight pipe is specified throughout, it will be impossible to find sizes that exactly fit the specified requirements. Select and specify the sizes nearest to the ideal solution. To develop a familiarity with the use of schedule numbers, show all specifications by schedule number. Figure 6.2 gives sufficient information for this purpose. Standard-weight pipe is classified as Schedule 40.

Solve this problem by reference to Fig. 6.7 using the information and instructions which follow:

Given Information:

1. Tank layout (Fig. 6.6) in elevation.
2. Tank A: Supplying tank with constant level supply.
3. Tank B: Mix tank, 600 cu. ft. vol.
4. Tank C: Mix tank, 2,400 cu. ft. vol.
5. Tank D: Mix tank, 3,000 cu. ft. vol.
6. Pipe to tank A, 4″ steel, Schedule 40.
7. Tanks B, C, and D to fill by gravity.

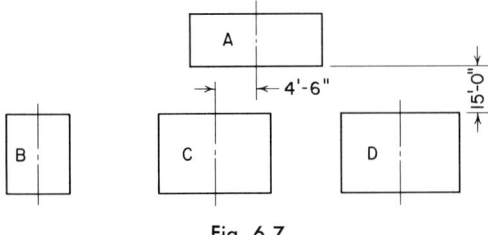

Fig. 6.7.

Instructions:

1. Use 11 x 17 sheet with standard layout.
2. Figure pipe sizes so that all tanks will fill at approximately the same time.
3. Disregard friction loss in pipe.
4. Draw a diagram and specify all pipe and fittings. (Show specifications on diagram adjacent to part specified.)
5. Show valves to control flow to any combination of tanks.
6. Locate all outlets and inlets on the center lines of the tanks.
7. Do not dimension.
8. Use single-line representation, with screwed fittings on all pipe under 2½″.
9. Specify all pipe by schedule number.
10. Show a sufficient number of unions to make disassembly convenient.
11. Select manually controlled valves that will offer the least restriction to flow.

7 Specification of parts

A draftsman is frequently called upon to make up material lists from specifications made up by the engineer, or from previously designed systems of pipe. Such a task requires not only the ability to read piping drawings, but also a familiarity with the fittings and materials available, and with the requirements of the system.

SELECTION OF MATERIALS

Catalog descriptions usually supply the information necessary for a part specification. It is not possible to discuss here all the valves and fittings available. When a fitting or valve is needed, the designer should acquaint himself thoroughly with the requirements, and with these requirements in mind, select from catalogs or other sources of information the valves or fittings that best meet the requirements.

Some of the points that should be considered when selecting fittings or valves are:

1. Pressure involved.
2. Corrosion.
3. Heat.
4. Strain to which the part is to be subjected.
5. Ease of installation.
6. Frequency of dismantling.
7. Ease and convenience of operation.
8. Cost.

SPECIFYING FITTINGS

Fittings should be specified by first stating the size and then the name of the fitting. The size is the same as the nominal size of the pipe used with the fitting. When a fitting connects more than one size pipe, the largest run opening is given first, followed by the opening size on the opposite end of the run. Side openings are given next, the larger being mentioned first.

The material of which the fitting is made should follow the name. For example, a cast-iron tee would be described as follows: 3 x 2 x 1, Flanged Tee, Cast Iron. Some draftsmen prefer to state a pressure rating, in which case the above specification would read: 3 x 2 x 1 Flanged Tee, Cast Iron, 125 lb.

If an external thread is wanted, the word "male" should follow the size designation of the opening where the external thread is located.

SPECIFYING VALVES

Valve specifications should include the nominal size of the valve, the pressure rating, the material in the body and trim, the valve description, and the type of connection. In actual practice, the name of the company manufacturing the valve may also be mentioned along with the catalog or figure number.

PART NO.	DESCRIPTION	MATERIAL	NO. REQ.

Fig. 7.1. Typical parts list form.

FIG. 7.2. PIPE HANGERS.
(Courtesy Grinnell Co.)

Fig. 7.2a.
Concrete insert.

Fig. 7.2b.
Concrete insert.

Fig. 7.2c.
Pipe hanger flange.

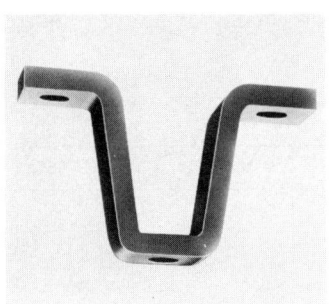

Fig. 7.2d.
Swivel hanger flange.

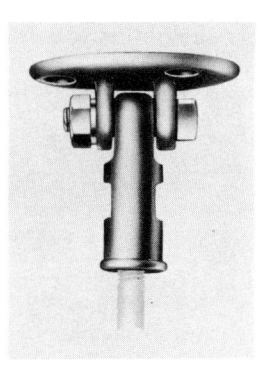

Fig. 7.2e.
Beam attachment.

THE PARTS LIST

The method of part specification varies from company to company. The draftsman must learn the method of specification used by the company for which he works.

A list of all required parts is usually made into a parts list, or material bill, which may appear on the main assembly drawing, or may be included separately with a set of drawings. This parts list should contain information needed for complete identification of the part.

If the part is numbered on the drawing, it should be numbered with the same number on the parts list. The parts list should also show the number of parts needed for the complete assembly together with the part name, size, material, and special information. A typical parts list form appears in Fig. 7.1.

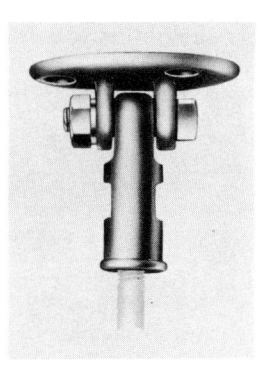

Fig. 7.2f.
Adjustable swinging hanger flange.

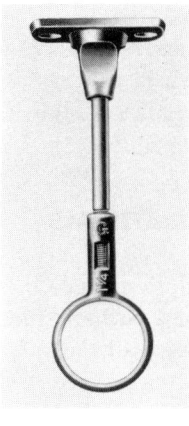

Fig. 7.2g.
Adjustable clip.

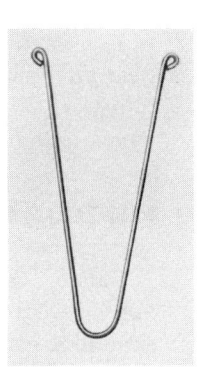

Fig. 7.2h.
U-hook.

ABBREVIATIONS*

There are many abbreviations common to pipe work which may appear in material lists or on drawings. Some of these abbreviations are listed below.

A. I.	All iron	B. P.	By-pass
AL	Aluminum	Br.	Brass
B. F.	Blind flange	B. & S.	Bell and spigot
B. M.	Brass mounted	B. W.	Butt weld

*For a more complete list of abbreviations it is suggested that the student refer to the following A.S.A. Publications: Abbreviations for Scientific and Engineering Terms (Z10.1-1941) and Abbreviations for Use on Drawings (Z32.13-1950).

SPECIFICATION OF PARTS

C. to F.	Center to face	I. B.	Iron body
C. I.	Cast iron	I. D.	Inside diameter
C. I. A.	Cast iron alloy	L.	Elbow
C. P.	Close pattern	L. H.	Left hand
C. S.	Cast steel	L. R.	Long radius
D. S.	Double sweep	L. S.	Lock shield or long sweep
E. to E.	End to end		
E. H.	Extra heavy	L. W.	Lap weld
F. & D.	Faced and drilled	Mall.	Malleable
		M. I.	Malleable iron
F. to F.	Face to face	N. R. S.	Nonrising stem
F. E.	Flanged ends	O. D.	Outside diameter
F. O.	Faced only	O. S. & Y.	Outside screw and yoke
F. S.	Forged steel		
F. W.	Full weight	P. E.	Plain end
Gal.	Galvanized	P. F.	Plain face
H. E.	Hub end	Q. O.	Quick opening

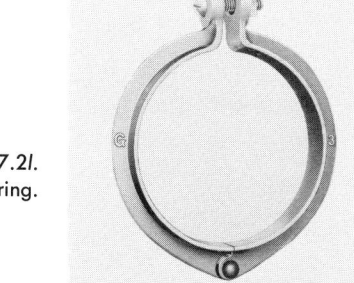

Fig. 7.2*l*.
Split pipe ring.

Fig. 7.2*m*.
Adjustable swivel ring.

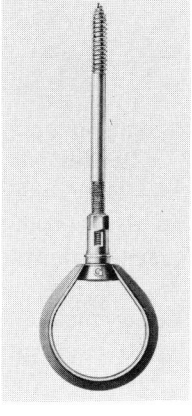

Fig. 7.2*i*.
Adjustable coach screw clip.

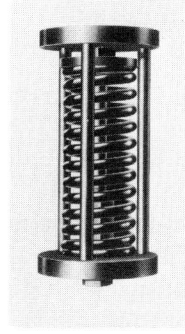

Fig. 7.2*j*.
Spring cushion pipe hanger.

Fig. 7.2*k*.
I-beam clamp.

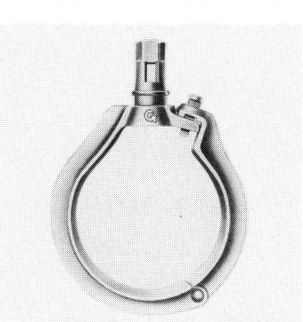

Fig. 7.2*n*.
Adjustable swivel split ring. ▶

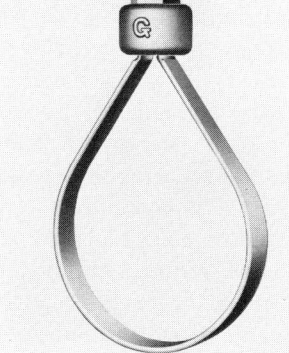

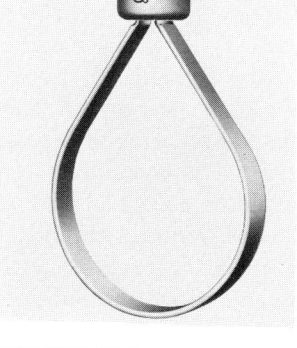

Fig. 7.2*o*.
Adjustable wrought pipe ring.

R. H.	Right hand	W. O. G.	Cold water or gas
R. & L.	Right and left		
S. E.	Screwed ends	X. H.	Extra heavy
S. O.	Side outlet	XX. H.	Double extra heavy
Std.	Standard		
T.	Tee	XX. S.	Double extra strong
W. I.	Wrought iron		

MISCELLANEOUS PARTS

A complete pipe drawing usually involves the use of many parts which are not actually a part of the piping system, but which are, nevertheless, essential

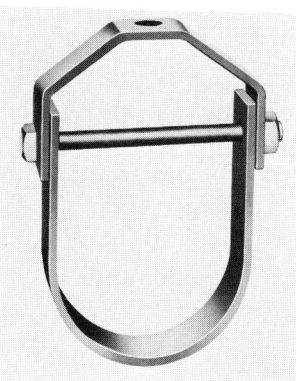

Fig. 7.2*p*.
Adjustable wrought clevis hanger.

FIG. 7.3. BRACKETS AND SUPPORTS.
(Courtesy Grinnell Co.)

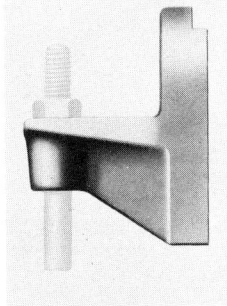

Fig. 7.3a. Side column bracket.

Fig. 7.3b. Side beam bracket.

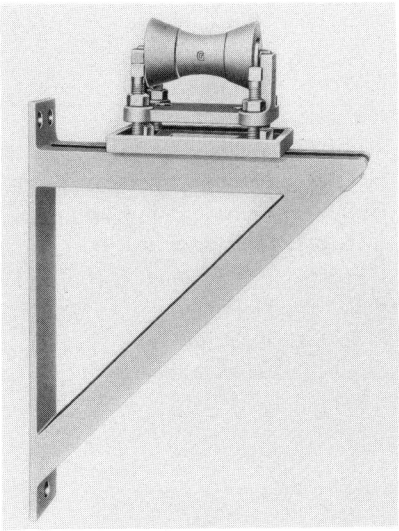

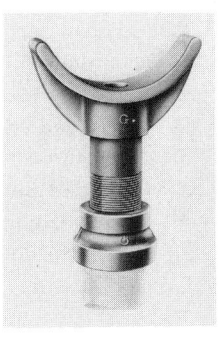

Fig. 7.3d. Adjustable pipe saddle support.

Fig. 7.3c. Welded steel bracket and pipe roll. ▶

and which require a place on the parts list as well as representation on the drawing.

CONCRETE INSERTS

If piping is to be placed in a building which is to be constructed or is under construction, the draftsman should give some thought as to the means of securing the pipe to the walls, ceiling, or floors. If there are no exposed structural members accessible for the securing of pipe hangers or supports, some type of concrete insert should be selected which could be placed as the concrete is poured. This, of course, requires careful planning on the part of the draftsman as well as cooperation between the piping engineer and the builder. If the piping plans are not ready at the time the concrete is poured, the inserts may be set at regular intervals. As is true in all piping problems, selection of inserts should be made with knowledge as to the usual practices of the company and with consideration as to the availability of the type of insert recommended. If the selection is up to the draftsman, he should refer to current catalogs and place his recommendations on a materials list in such a way as to give complete information for ordering by the purchasing agent for the company.

Figures 7.2a and 7.2b illustrate two common types of concrete inserts. Figures 7.2c to 7.2g illustrate other devices used for securing pipe to the ceiling. Figures 7.2h and 7.2i illustrate two simple, economical methods for hanging pipe to wooden supports. Figure 7.2k illustrates a method for hanging pipe to I-beams. There are many other devices on the market used for hanging pipe from overhead supports, or the draftsman may choose to design supports; if so, the details of the design should be laid out on the sheet and the materials needed listed on the materials list.

PIPE HANGERS

The term "pipe hangers" usually refers to the piece that suspends the pipe from the insert, flange, or clamp at the ceiling. Pipe hangers vary in complexity from single strips of iron or wire, as illustrated in Fig. 7.2h, to the spring-cushion-type hanger illustrated in Fig. 7.2j. Pipe is secured to the hanger by means of pipe rings, some of which are illustrated in Figs. 7.2l to 7.2p. Pipe may also be secured to the wall by means of brackets, three of which are illustrated in Figs. 7.3a to 7.3c. The pipe may also be supported from the floor or from some other desirable structure beneath the pipe by supports such

SPECIFICATION OF PARTS

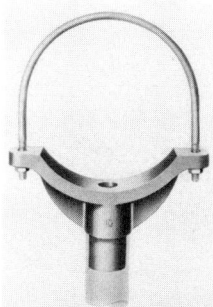

Fig. 7.3e. Pipe stanchion saddle.

Fig. 7.3f. Pipe covering protection saddle.

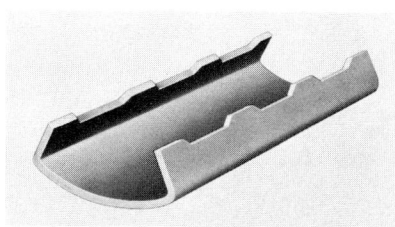

as the ones illustrated in Figs. 7.3d and 7.3e or by the roller support shown in Fig. 7.3c.

INSULATION

Insulation is used on pipe when problems of temperature are involved. The nature of the installation is usually stated for the draftsman, but it is important that the draftsman know something of the composition of the insulation so that he can make accurate layouts and adequate provisions for supports. The thickness of the insulation around the pipe varies according to the need and the size of the pipe. Since the insulation is soft, it is important that metal pieces called saddles be installed at the points of support. Figure 7.3f illustrates a saddle. This saddle is spot welded to the pipe as illustrated in Fig. 7.4. In the installation, dimensions are taken from the center of the pipe to the bottom of the saddle.

ANCHORS

In most installations, provision must be made for protection of machinery or equipment from strain caused by contraction and expansion of pipe. It is important to avoid any excessive stress on machinery or equipment caused by pipes having been too rigidly connected or by any other improper connections. Connections should never be made at sharp angles or so that pipe will push against the equipment. They must be made so that any movement will be absorbed by the pipe and not by the machinery. If there is danger of strain on the machinery, the pipe should be firmly anchored at a point close to the machine. The anchor may be a concrete island on the floor, or it may be fabricated and secured to the wall, floor, or ceiling. Piping plans should include details describing the construction of anchors, and materials required should be included in the materials list.

FABRICATED PIPING

In addition to the standard parts in a system, special fabricated piping may be utilized. The company employing the draftsman may find it practical to order certain fabricated piping such as expansion loops, offset bends, or piping designed to special specifications. Any special part should be noted on the drawing and listed in the parts list. If the part in question is to be fabricated in the company shop, this should be noted, with reference to the detail drawing on which the part is described.

A more complete discussion relative to fabricated piping may be found in Chapter 9.

Problem

Problem 1. The accompanying drawing (Fig. 7.5) is a piping layout for a unit of machinery. Two views are shown as they would appear in the plan and elevation of a building.

Draw a single-line developed view of the system showing all valves and fittings in plan.

In the lower right-hand corner of the sheet above the title block show a material bill that will include specifications and number required for all fittings. If necessary, refer to illustrations in this text by figure number. (Normally catalog descriptions would be used.)

All pipe is to be 1½", Schedule 40 steel, except that to the gage which is to be ½".

The quantity of pipe need not be estimated.

The drawing may be done in pencil on tracing paper. Sheet size 11 x 17.

Note: It is common practice, when a symbol representing a valve or other detail appears in one view, to omit it in other views, especially when lines coincide as when one line appears directly above another.

FUNDAMENTALS OF PIPE DRAFTING

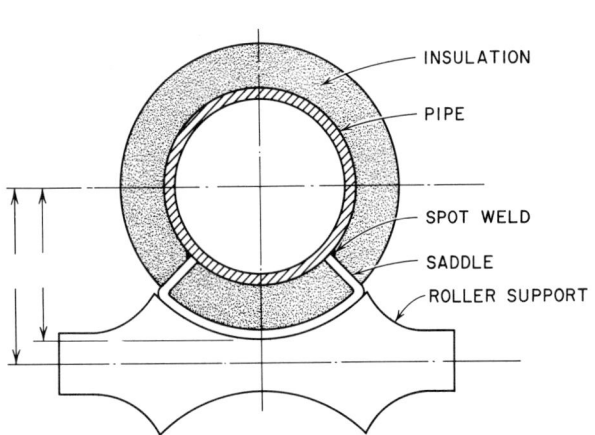

Fig. 7.4. Section through insulated pipe and saddle.

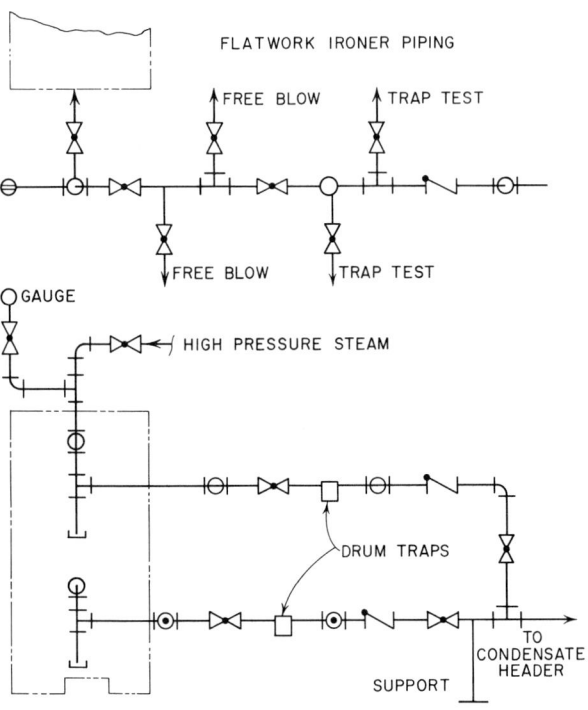

Fig. 7.5. Flatwork ironer piping.

8 General arrangement and diagram drawings

The first consideration to be given to a piping layout is to the plot plan which gives the location of all piping installations of the plant.

The second consideration of any over-all piping design is the arrangement of the machines and piping within each individual unit or building. One of the first considerations here would be information relative to the location of the points of connection with other installations of the plant.

GENERAL ARRANGEMENT DRAWING

A drawing that shows the location of machinery and its connecting pipes within an installation is called a general arrangement drawing. This type of drawing is usually made to a larger scale than a plot plan and has sufficient detail to give complete information about the location of pipes, machinery, and equipment. It may also contain information about the structural features of the building.

In simple installations where pipe sizes are small and the parts standard, such as the water piping in a private dwelling, a general arrangement drawing may supply all the information necessary for the workman to make a complete installation. The pipe, being small, would be cut to length and threaded by the worker making the installation. Since the units are more or less standard in their connections, the workman could do the necessary planning without complicated instructions.

If the building or assembly area is limited, a simplified drawing showing machinery and equipment in outline form serves as an aid for studying arrangements of machinery and clearances for pipe, or as a key sheet for further, more detailed drawings. If the size of the building or area is undecided, a general arrangement drawing may serve as a preliminary study for building design.

General arrangement drawings may vary from simple single-line diagrams with only a minimum of details showing, to double-line drawings showing practically all details. It is common practice to make a simple single-line layout upon which the location of machines and piping is shown. The more complicated details are then identified by letter, number, or name, and redrawn to a larger scale. This detail drawing may appear on the same sheet with the layout, or may be drawn on a separate sheet or sheets. It is important that details be well and clearly labeled so that they can be easily identified on the general arrangement drawing.

Frequently, single-line isometric or pictorial drawings of portions of the general arrangement drawing help in the understanding of the over-all installation. These pictorial drawings may include information about pipe sizes, type of fittings, valve types, or any other information which may add to the clarity and completeness of the drawing.

Every installation brings up new problems for the draftsman. It should be remembered that first in importance is an accurate knowledge of the problem at hand. No definite set procedure can be recommended, for in the final summation, the clarity of the drawing depends upon the judgment and ingenuity of the draftsman.

DIMENSIONING

A general arrangement drawing should show center-to-center dimensions of all pipe as well as locating all valves, traps, strainers, etc. by stating the location of the center of the part in question in relation to some other stated part or center line. The nominal pipe sizes are usually written beside the pipe. On some drawings, the fitting description and size are omitted, leaving that information to be obtained by reading the symbols and by observing pipe sizes to the fittings. If room allows, confusion on the part of the reader can be avoided by labeling each fitting with its proper description and size. Valves and other equipment are frequently identified by the manufacturer's number on the drawing. Over-all dimensions of valves and fittings are seldom shown, since their sizes are standard; however, on detail drawings, it may be necessary to state dimensions of standard parts in order to scale crowded parts accurately.

The draftsman should be careful to give enough location dimensions that the piping can be accurately located in relation to the parts of the room. It is good practice, when possible, to select one side and one end of a room as base lines and indicate all location dimensions from these lines.

Vertical dimensions are often established by reference to a pre-established base line of a known or assumed elevation. When a base line of this type is used, all piping in the entire building or system can be dimensioned so as to show the elevation of all lines and equipment in their correct relative elevations.

General arrangement drawings may be used as flow charts to show heat balance or load, or as key drawings to show locations of parts or subinstallations detailed on other sheets.

General arrangement drawings should be made with close reference to machinery and equipment drawings (usually supplied by the manufacturer of the equipment and machinery), which give detailed information about the points of attachment of the pipe, the size of the equipment, the clearances required, and any other information that may deal with the proper installation and operation of the equipment. Figures 5.7a and 5.7b illustrate drawings of this type.

APPEARANCE

As much as possible, the piping should follow the lines of the building. Lines running in the same direction should be parallel, and a jumbled, confused appearance avoided as much as possible.

CLEARANCE

Free space around a pipe is called "clearance." Clearance can best be provided by making sure that a pipe is allowed to continue its run in the same horizontal or vertical plane. If this is always taken into consideration, other pipes may be run in the same planes as the first, providing a neat, organized appearance. This also will provide for easier access to each individual pipe in the event of inspection or disassembly.

Spaces between pipes should be large enough to allow for the use of a wrench in the assembly or disassembly, for insulation, if it is used, or to provide ample room for flanges or other connections or valves should they appear side by side.

Abrupt changes in direction should be avoided as much as possible to avoid the formation of pockets in the piping which might interfere with the flow.

Clearance around the pipe in relation to features of the building should also be considered. Pipe should not be hung so close to a wall or ceiling as to make installation difficult.

SUPPORT

The third matter of importance in arranging pipes is that of support. The structural features of the building should be strong enough to support the pipe lines, and the supports of the pipe itself should be spaced closely enough to avoid sagging of the pipe. On pipe lines of 2½-inch pipe and smaller no more than 10 feet is considered good spacing. On pipe up to 6 inches nominal size, hangers may be spaced at 12 feet. For pipe sizes above 6 inches, hanger spacing should usually be no more than 15 feet.

Many times the ideal arrangement for pipe would be an overhead assembly, but if the roof members are not strong enough to support both the roof and the pipe lines, other arrangements must, out of necessity, be made. It is common practice to require that the supporting members should be capable of supporting ten times as much weight as they will be called upon to support.

OTHER FACTORS

Other factors, such as expansion and contraction and adequate anchors to prevent damage to equip-

GENERAL ARRANGEMENT AND DIAGRAM DRAWINGS

ment, as well as arrangement for convenient and safe operation should be given careful consideration.

Still another feature to consider in the general arrangement drawing is the supports for machinery in the installation. This necessitates a thorough knowledge of the building design. Footings should be provided for heavy machinery. Frequently machinery may be installed on concrete islands.

Problem

Problem 1. Commercial laundries, hospitals, colleges, hotels, and other institutions commonly use the washer piping hookup illustrated in Fig. 8.1. It provides the usual hot and cold water and steam connections to a battery of washers. Although actual installations may

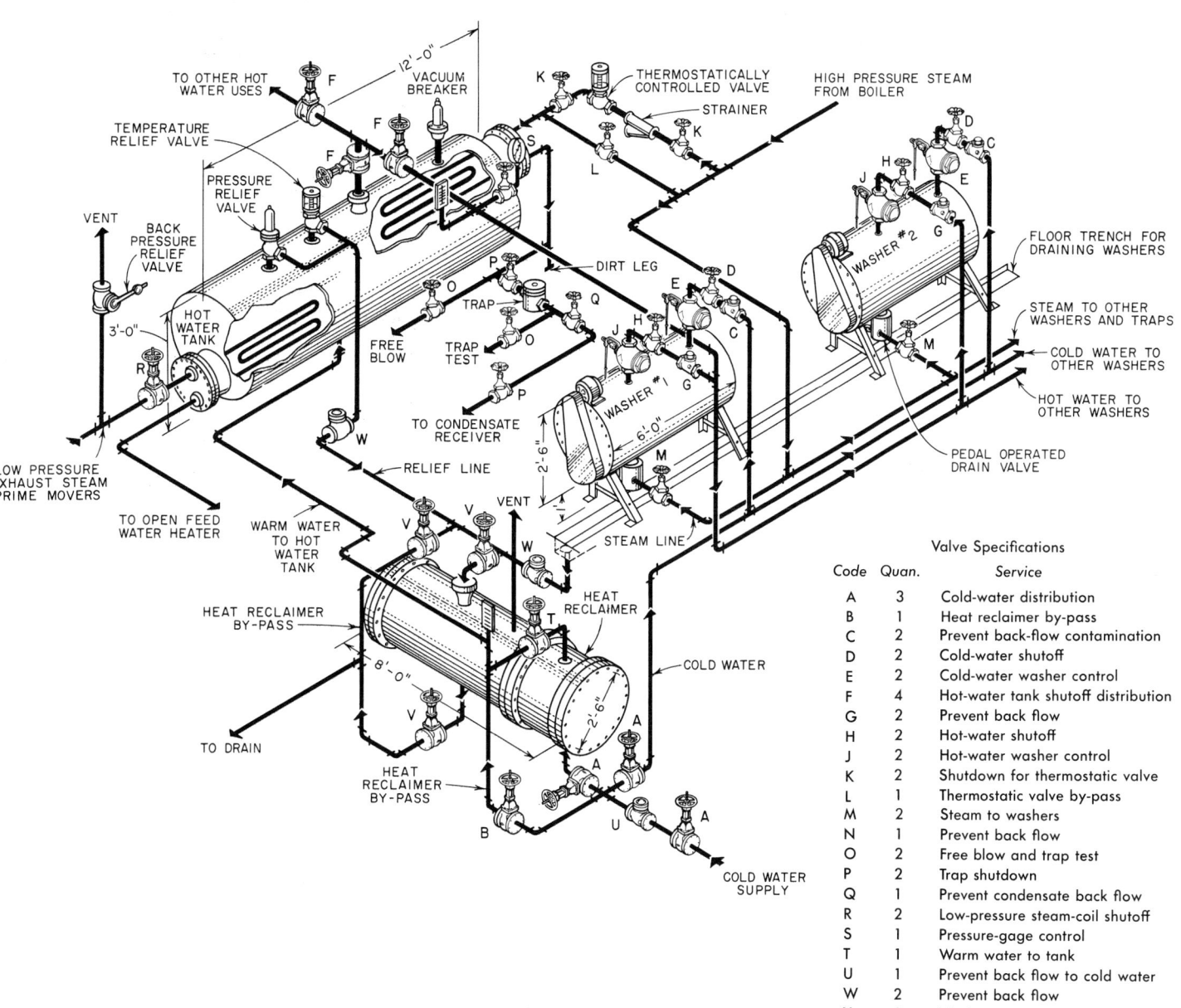

Fig. 8.1. Commercial laundry washer piping. (Courtesy Jenkins Bros.)

differ widely in complexity, size, and method of operation, this diagram shows control points where valves are essential to any soundly engineered installation.

Waste from washers drains into a common trench, which in turn drains into a heat reclaimer. Cold water that is to be heated first passes through a series of coils in the heat reclaimer, where it is preheated. This utilization of heat from the hot wastes before they are discharged to the sewers improves plant efficiency.

The hot-water tank is fitted with two heating coils. The primary coil is supplied with the low-pressure exhaust steam from the various steam power units such as pumps, engines, and turbines. This coil is generally sufficient to heat water for the periods of operation when the hot-water requirements are low. An auxiliary heating coil supplied directly with high-pressure steam from the boiler is thermostatically controlled to heat the water during peak demands.

A rapid-action globe valve with catch-open lever is recommended for use in laundry and other services where quick on-and-off action is required.

Using the isometric layout of Fig. 8.1 as a guide, draw the laundry piping and equipment for the building shown in Fig. 8.2.

Instructions:

1. Use simplified double-line symbols.
2. Select a sheet size that will permit the use of a scale satisfactory for double-line drawing. If necessary, use more than one sheet.
3. Work on tracing paper with pencil.
4. Show plan and elevation with hot-water tank, heat reclaimer, and two washers in position, with locations shown in relation to building walls. Also, locate and dimension all points where pipe enters the room.
5. For the sake of simplicity, all pipe sizes will be 2″.
6. All hangers and supports are to be indicated. Consider weight of pipe and supporting features of the rooms.
7. Select correct valves and locate with respect to accessability.
8. Make a materials bill showing the number required and the specifications for all fittings, valves, hangers, and supports.

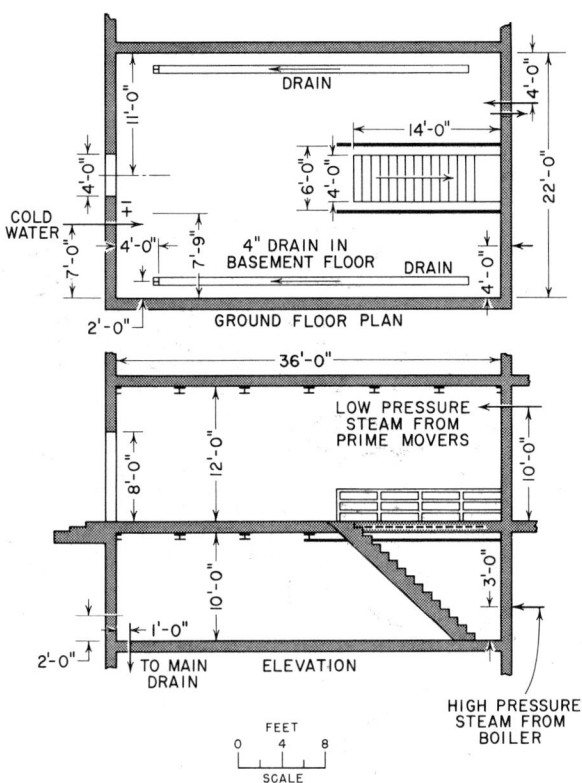

Fig. 8.2.

9 Detail drawing

The term "detail drawing" suggests to the piping draftsman drawings made for several different purposes and related to the completed set of piping plans in different ways. For example, a general piping layout may, on the same sheet, show details of certain parts of the drawing where clarification is needed, or explanatory details may be drawn up on single sheets accompanying a set of piping drawings. These details may be of single parts or of assembled parts. In any case, detail drawings should be drawn to a scale large enough to show the work adequately, probably between $1'' = 1'$-$0''$ and $3'' = 1'$-$0''$.

SHOP DRAWING

The term "shop drawing" is usually used to refer to drawings made for use in the fabrication shop. It is a detail drawing which gives all the necessary information for the fabrication of a part. For example, a 90° bend may appear on the assembly without detailed information regarding the space needed for assembly, the radius of the bend, identification of the parts used, or other information. The shop drawing will give all information needed by the shop man for making a part.

Many companies follow the practice of fabricating for future use the common standard types of pipe bends. If the plant is not equipped for fabrication, then standard bends can be purchased. Figure 9.1 illustrates the standard types of flanged-end pipe bends commonly listed in manufacturer's catalogs. Any type of end connection may, of course, be specified and obtained.

Whether the company makes or orders the needed part, a detailed drawing should be available, as an aid either in the fabrication or in the specification for ordering; otherwise accurate space and clearance allowances cannot be calculated. In some cases, catalog drawings and descriptions may be referred to, thus avoiding the necessity of making detail drawings.

The draftsman should be concerned not only with detailed information regarding single fabricated parts, but also with how all parts fit together in an assembly or subassembly.

ADVANTAGES OF SHOP FABRICATION AND ASSEMBLY

Assembly or fabrication in the plant shop has many advantages over field fabrication or assembly. In the first place, if the steps are properly planned, parts can be fabricated or assembled with much more efficiency than in the field because operating conditions can be much better controlled. Tools and machines that are too big or too intricate to handle in the field can be used to advantage in the shop. Also, skilled labor can be concentrated at a point where skill is most needed, and is enabled to work under conditions most conducive to quality work. Of course, work that is done in the shop for use in the field demands careful planning by experienced planners. The draftsman who takes part in this planning must be thoroughly familiar, not only with the processes involved and with the requirements of every situation, but also with the materials and equipment available. The draftsman who can draw accurate and dependable plans for the shop

FUNDAMENTALS OF PIPE DRAFTING

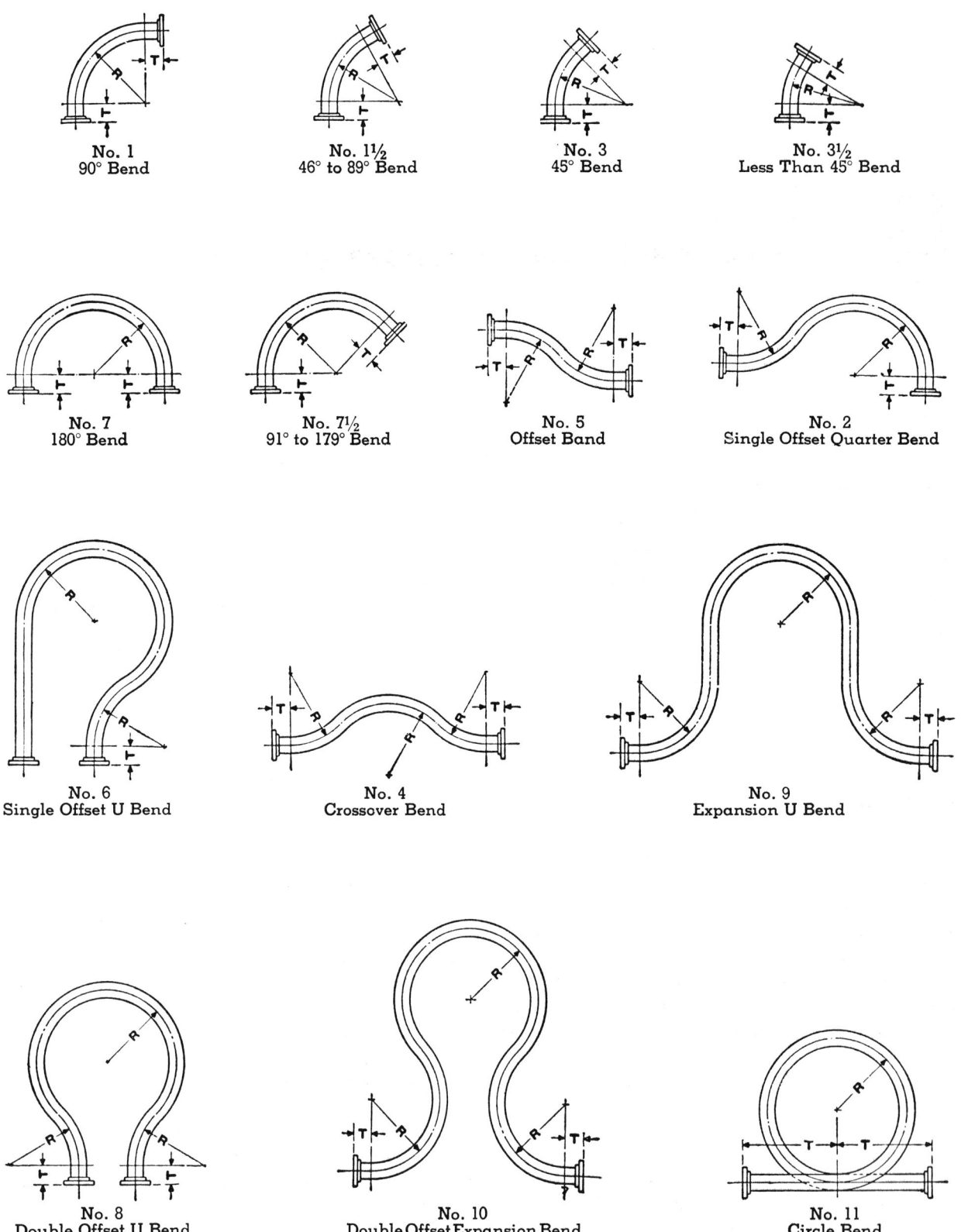

Fig. 9.1. Standard types of pipe bends. (Courtesy Walworth Co.)

DETAIL DRAWING

can save his employer many dollars in time, equipment, and material.

What the Draftsman Should Know

PARTS AND THEIR USES

In addition to being skilled in the use of drafting equipment and well informed on the theory of drafting, the piping draftsman must be familiar with the thousands of parts used in piping and capable of selecting these parts according to the needs of the design. Although most parts used in piping have standard dimensions, the draftsman should know how and where to find these dimensions in catalogs and descriptive material used by the company for which he works. He should know how to describe purchased parts properly.

PROBLEMS OF ASSEMBLY

Second in importance to the selection of the right part is the ability to make the necessary calculations for fitting the part into its proper place. A screwed valve, for example, although it requires a certain specified space in the line, also requires a certain added length to the pipe to provide for the distance the pipe enters the valve. This length varies with the size of pipe. Figure 9.2 illustrates standard thread lengths for different-sized pipe, and Fig. 9.3 lists the normal engagement of the pipe with the valve or mating part for a tight fit. Pipe used with flanged fittings may also require certain adjustments to allow for gaskets between flanges. There are, of course, many other points to be taken into consideration when fabricating or assembling pipe. The following paragraphs discuss in more detail the considerations involved in pipe detailing.

PIPE THREADS

Pipe 2½ inches and under which will be used under standard pressure and low temperatures is commonly assembled using threaded fittings and valves. (This practice will be discussed further in the discussion on welding.)

The American Standards Association (ASA) provides two types of threads for use on pipe. The normal type employs a taper internal and a taper external thread. The taper employed is $\frac{1}{16}$ inch per foot measured on the diameter. Figure 9.2 illustrates the characteristics of the thread. The table in Fig. 9.2 shows the standard dimensions for nominal pipe sizes. Figure 9.4 is the ASA recommendation for thread measurements for mechanical joints. It is common practice to use taper threads for external members and straight threads for internal members on the assumption that the materials are sufficiently ductile to allow the threads to adjust themselves to the taper thread.

The ASA recommends taper threads for all except the following five types of joints, which may utilize straight threads:

TYPE 1. Pressure-tight joints for pipe couplings.
TYPE 2. Pressure-tight joints for grease-cup, fuel, and oil fittings.
TYPE 3. Free-fitting mechanical joints for fixtures.
TYPE 4. Loose-fitting mechanical joints with lock nuts.
TYPE 5. Loose-fitting mechanical joints for hose couplings.

The number of threads per inch is the same in taper and straight pipe threads.

It should be observed from the table of Fig. 9.3 that the normal thread engagement of the taper thread varies with the size of the pipe. This normal thread engagement must be taken into consideration when drawing pipe assemblies that are to meet certain specified measurements.

The ASA recommendation for the representation of pipe threads is the same as for all other threads, the regular or simplified form of representation being acceptable. Some draftsmen prefer to show an exaggerated taper when taper threads are drawn. Regardless of the form used, the draftsman should be consistent in his method of representation. Figure 9.5 illustrates the conventional methods of representing pipe threads.

When a complete thread description is required it is given by stating the nominal pipe diameter, the number of threads per inch, and the standard letter symbol to denote the type of thread.

The following abbreviations are recommended by the ASA for specifying thread types:

NPT   Taper pipe thread
NPTF   Taper pipe thread (dryseal)
NPS   Straight pipe thread

FUNDAMENTALS OF PIPE DRAFTING

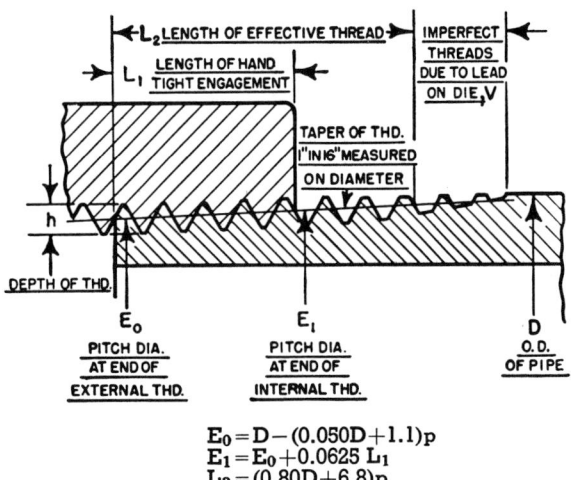

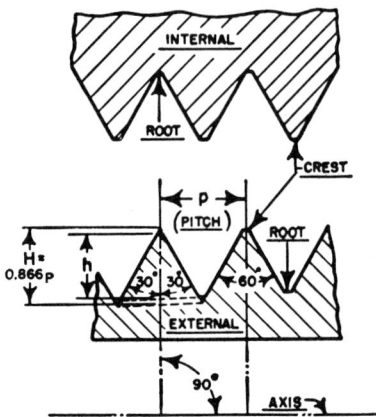

$E_0 = D - (0.050D + 1.1)p$
$E_1 = E_0 + 0.0625 L_1$
$L_2 = (0.80D + 6.8)p$
$H = 0.866p$

Nominal Pipe Size Inches	D	No. Threads per inch N	E_0	E_1	L_2	L_1	V	h	Pitch of Thread p
1/8	.405	27	.36351	.37476	.2639	.180	0.1285	.02963	.03704
1/4	.540	18	.47739	.48989	.4018	.200	0.1928	.04444	.05556
3/8	.675	18	.61201	.62701	.4078	.240	0.1928	.04444	.05556
1/2	.840	14	.75843	.77843	.5337	.320	0.2478	.05714	.07143
3/4	1.050	14	.96768	.98887	.5457	.339	0.2478	.05714	.07143
1	1.315	11½	1.21363	1.23863	.6828	.400	0.3017	.06957	.08696
1¼	1.660	11½	1.55713	1.58338	.7068	.420	0.3017	.06957	.08696
1½	1.900	11½	1.79609	1.82234	.7235	.420	0.3017	06957	.08696
2	2.375	11½	2.26902	2.29627	.7565	.436	0.3017	.06957	.08696
2½	2.875	8	2.71953	2.76216	1.1375	.682	0.4337	.10000	.12500
3	3.500	8	3.34062	3.38850	1.2000	.766	0.4337	.10000	.12500
3½	4.000	8	3.83750	3.88881	1.2500	.821	0.4337	.10000	.12500
4	4.500	8	4.33438	4.38712	1.3000	.844	0.4337	.10000	.12500
5	5.563	8	5.39073	5.44929	1.4063	.937	0.4337	.10000	12500
6	6.625	8	6.44609	6.50597	1.5125	.958	0.4337	.10000	.12500
8	8.625	8	8.43359	8.50003	1.7125	1.063	0.4337	.10000	.12500
10	10.750	8	10.54531	10.62094	1.9250	1.210	0.4337	.10000	.12500
12	12.750	8	12.53281	12.61781	2.1250	1.360	0.4337	.10000	.12500
14	14.000	8	13.77500	13.87262	2.2500	1.562	0.4337	.10000	.12500
16	16.000	8	15.76250	15.87575	2.4500	1.812	0.4337	.10000	.12500
18	18.000	8	17.75000	17.87500	2.6500	2.000	0.4337	.10000	.12500
20	20.000	8	19.73750	19.87031	2.8500	2.125	0.4337	.10000	.12500
24	24.000	8	23.71250	23.86094	3.2500	2.375	0.4337	.10000	.12500

Fig. 9.2. American Standard taper pipe threads. (Data from *Pipe Threads*, ASA B 2.1–1945, published by American Society of Mechanical Engineers, New York.)

DETAIL DRAWING

NPSC   Straight pipe thread in couplings
NPSI   Intermediate internal straight pipe thread (dryseal)
NPSF   Internal straight pipe thread (dryseal)
NPSM   Straight pipe thread for mechanical joints
NPSL   Straight pipe thread for locknuts and locknut pipe thread
NPSH   Straight pipe thread for hose couplings and nipples
NPTR   Taper pipe thread for rail fittings

Example: 2½–8 NPT
Explanation: 2½ = Nominal pipe diameter
         8 = Number of threads per inch
         N = American Standard Thread
         P = Pipe
         T = Taper

It is common practice to shorten the above description as follows: 2½ NPT.

FLANGES

Much industrial piping makes use of flanged valves and fittings. Ease of assembly and disassembly is an important feature in favor of the use of flanged joints. For example, large pipe made up of screwed joints would require large wrenches and possibly other equipment for assembly, and the use of flanged joints for the same size would involve the use of comparatively light equipment. The use of flanges in a piping system also permits the removal of valves or other equipment in the assembly without disturbing the rest of the assembly.

Figures 6.4 and 6.5 illustrate the common types of flanges and flange facings. Figure 9.6 is a duplication of a page from a manufacturer's catalog.

All dimensions of flanges, including the bolting, are standardized; however, there is a considerable variation in the types of facing and method of attaching flanges to pipe. A flange may be threaded (Fig. 6.4d) to provide for threaded pipe (threaded flanges are frequently welded to the pipe at the back of the flange), or it may be bored to allow the pipe to pass through the flange. The slip-on flange is of the latter type. It allows for welding at both the face and back of the flange (Fig. 6.4a). The lap-joint flange (Figs. 6.4c and 6.5b) provides for the flaring or the upsetting of the pipe on the face side of the flange. The

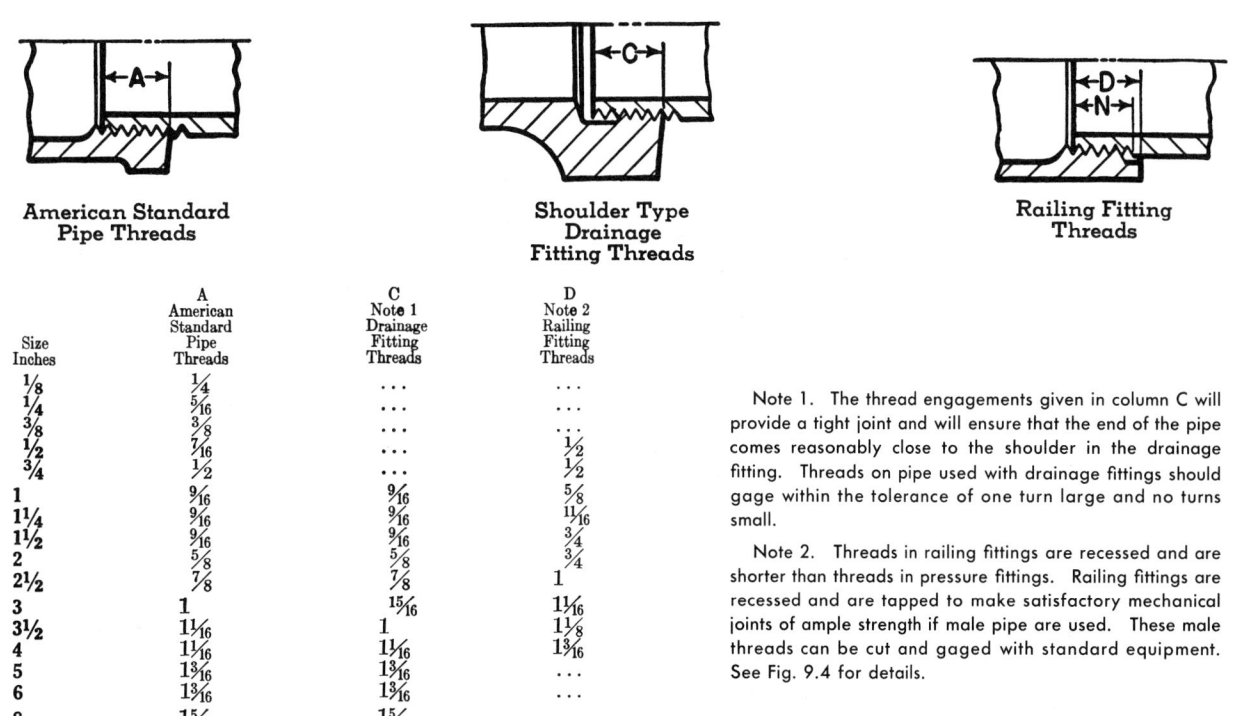

Size Inches	A American Standard Pipe Threads	C Note 1 Drainage Fitting Threads	D Note 2 Railing Fitting Threads
⅛	¼	...	...
¼	⁵⁄₁₆	...	...
⅜	⅜	...	...
½	⁷⁄₁₆	...	½
¾	½	...	½
1	⁹⁄₁₆	⁹⁄₁₆	⅝
1¼	⁹⁄₁₆	⁹⁄₁₆	¹¹⁄₁₆
1½	⁹⁄₁₆	⁹⁄₁₆	¾
2	⅝	⅝	¾
2½	⅞	⅞	1
3	1	¹⁵⁄₁₆	1¹⁄₁₆
3½	1¹⁄₁₆	1	1⅛
4	1¹⁄₁₆	1¹⁄₁₆	1³⁄₁₆
5	1³⁄₁₆	1³⁄₁₆	...
6	1³⁄₁₆	1³⁄₁₆	...
8	1⁵⁄₁₆	1⁵⁄₁₆	...
10	1½	1½	...
12	1⅝	1⅝	...

Note 1. The thread engagements given in column C will provide a tight joint and will ensure that the end of the pipe comes reasonably close to the shoulder in the drainage fitting. Threads on pipe used with drainage fittings should gage within the tolerance of one turn large and no turns small.

Note 2. Threads in railing fittings are recessed and are shorter than threads in pressure fittings. Railing fittings are recessed and are tapped to make satisfactory mechanical joints of ample strength if male pipe are used. These male threads can be cut and gaged with standard equipment. See Fig. 9.4 for details.

Fig. 9.3. Normal thread engagement necessary to make a tight joint. Dimensions given in the table may vary due to shrinkage of castings, tapping tolerance, etc. (Data from *Pipe Threads*, ASA B 2.1–1945, published by American Society of Mechanical Engineers, New York.)

FUNDAMENTALS OF PIPE DRAFTING

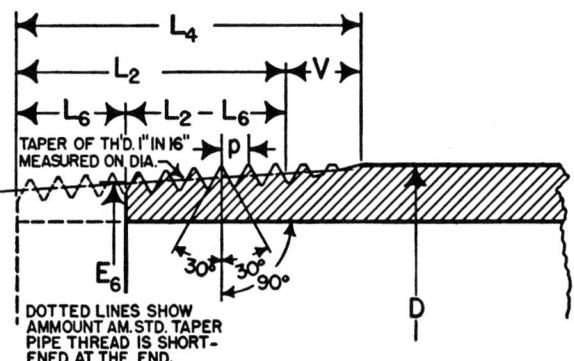

Nominal Pipe Size Inches	D Outside Diameter of Pipe Inches	Number of Threads per Inch	Depth of Thread Inches	E₆ Pitch Diameter at End of External Thread Inches	L₂-L₆ Length of Effective Thread of Am. Std. Pipe Thds. Inches	L₄-L₆ Total Length of External Thread, Inches Max.	L₆ Shortening of Am. Std. Pipe Thread		V Imperfect Threads Due to Lead of Die Max.	
							Thds.	In.	Thds.	In.
½	0.840	14	0.0571	0.7718	0.320	0.499	3	0.214	2½	0.179
¾	1.050	14	0.0571	0.9811	0.332	0.510	3	0.214	2½	0.179
1	1.315	11½	0.0696	1.2299	0.422	0.639	3	0.261	2½	0.217
1¼	1.660	11½	0.0696	1.5734	0.446	0.707	3	0.261	3	0.261
1½	1.900	11½	0.0696	1.8124	0.463	0.724	3	0.261	3	0.261
2	2.375	11½	0.0696	2.2853	0.496	0.757	3	0.261	3	0.261
2½	2.875	8	0.1000	2.7508	0.638	1.013	4	0.500	3	0.375
3	3.500	8	0.1000	3.3719	0.700	1.075	4	0.500	3	0.375
3½	4.000	8	0.1000	3.8688	0.750	1.125	4	0.500	3	0.375
4	4.500	8	0.1000	4.3656	0.800	1.175	4	0.500	3	0.375

Fig. 9.4. Dimensions of external taper pipe threads for railing fittings (mechanical joints). (Dimensions given in the table agree with those in *Pipe Threads*, ASA B 2.1–1945, published by American Society of Mechanical Engineers, New York.)

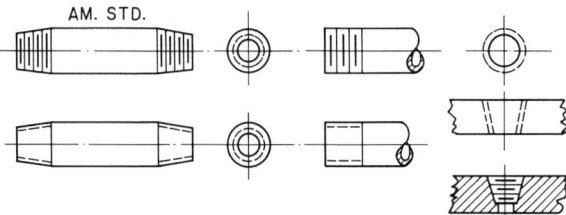

Fig. 9.5. Conventional representation of pipe threads.

welding-neck flange (Fig. 6.4b) provides a long chamfered neck suitable for attaching the pipe by welding. The welding-neck flange is probably the most commonly used of all flanges.

The threaded flange is used mainly for two purposes at the present time: (1) for repair work in areas where welding is hazardous, and (2) in high-pressure piping where the metal in the pipe is not suitable for welding. As previously stated, threaded flanges are frequently seal welded, but this is not entirely satisfactory and should be avoided when possible.

The method of attachment should be given careful consideration when detailing pipe and estimating pipe lengths, otherwise space allowances will be inaccurate.

Flanges are assembled by inserting a gasket between the faces of the flanges, which are then drawn tightly together by bolts. This gasket usually occupies a space of ¹⁄₁₆ inch. In some flanges, the seal is accomplished by compressing a sealing ring in a circular ring groove in each mating flange (Fig. 6.5f) as the flanges are drawn together by bolts. When the ring seal is used, the space between faces varies between ⁵⁄₃₂ inch and ⁷⁄₁₆ inch depending upon the size of the flange and the pressure involved.

Most steel flanges have raised faces, as illustrated in Fig. 9.6. Cast-iron flanges are commonly made with flat faces on the full face of the flange, except the 250-pound flange which has a ¹⁄₁₆-inch raised face.

Since flange measurements are standardized, the draftsman should not attempt to design or detail work involving flanges without having available accurate information regarding these standard measurements, as well as other pertinent information such as the type of

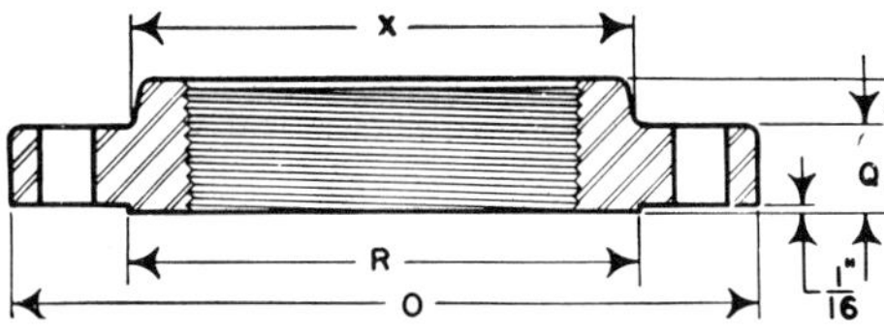

Threaded

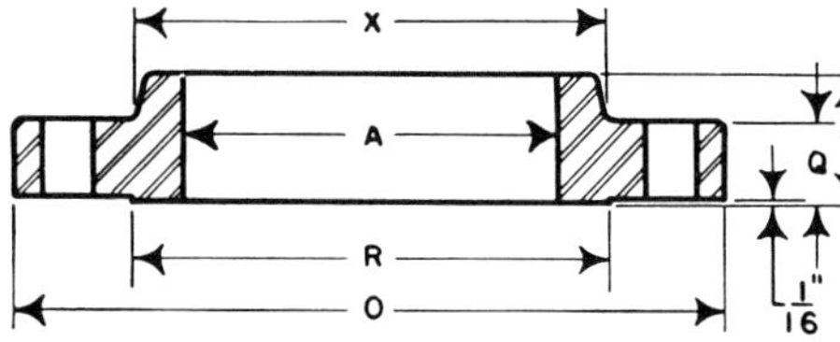

Slip-On

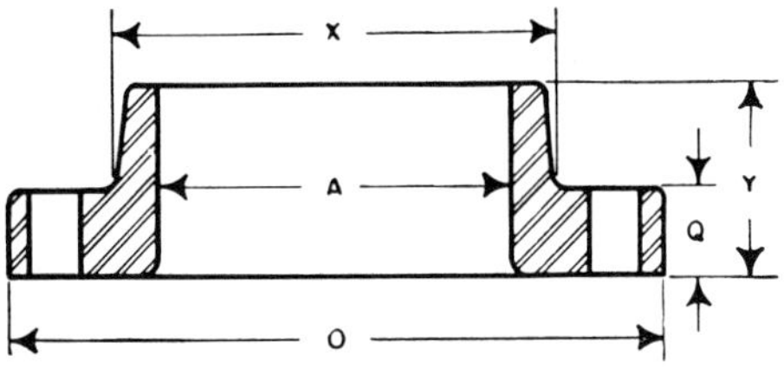

Lap

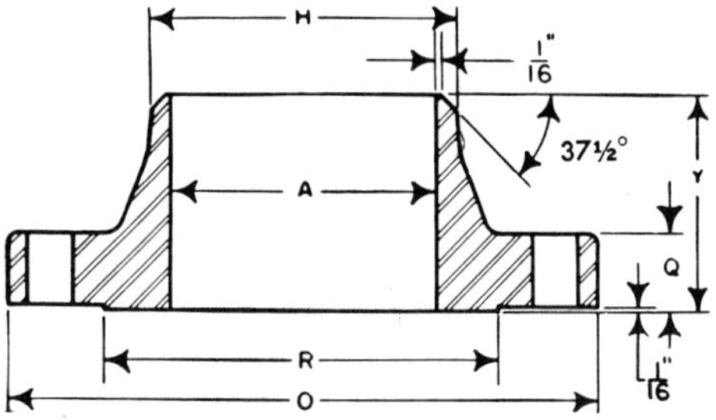

Welding Neck

	Dimensions, Inches													
						Y — Length through Hub			A — Diameter of Bore					
Nominal Pipe Size Inches	O Outside Diameter of Flange	††Q Thick. of Flange Minimum	‡‡R Outside Diameter of Raised Face	X Hub Diameter at Base	*H Hub Diam. at Beginning of Chamfer	Threaded, Slip-On	Lap	Welding Neck	Slip-On	Lap	**Welding Neck	No. of Bolts	Size of Bolts Inches	Diam. of Bolt Circle Inches
1/2	3 1/2	7/16	1 3/8	1 3/16	.84	5/8	5/8	1 7/8	▲	31/32	5/8	4	1/2	2 3/8
3/4	3 7/8	1/2	1 11/16	1 1/2	1.05	5/8	5/8	2 1/16	▲	1 3/16	13/16	4	1/2	2 3/4
1	4 1/4	9/16	2	1 15/16	1.32	11/16	11/16	2 3/16	▲	1 7/16	1 1/16	4	1/2	3 1/8
1 1/4	4 5/8	5/8	2 1/2	2 5/16	1.66	13/16	13/16	2 1/4	▲	1 25/32	1 3/8	4	1/2	3 1/2
1 1/2	5	11/16	2 7/8	2 9/16	1.90	7/8	7/8	2 7/16	▲	2 1/64	1 5/8	4	1/2	3 7/8
2	6	3/4	3 5/8	3 1/16	2.38	1	1	2 1/2	▲	2 1/2	2 1/16	4	5/8	4 3/4
2 1/2	7	7/8	4 1/8	3 9/16	2.88	1 1/8	1 1/8	2 3/4	▲	3	2 15/32	4	5/8	5 1/2
3	7 1/2	15/16	5	4 1/4	3.50	1 3/16	1 3/16	2 3/4	▲	3 5/8	3 1/16	4	5/8	6
3 1/2	8 1/2	15/16	5 1/2	4 13/16	4.00	1 1/4	1 1/4	2 13/16	▲	4 1/8	3 17/32	8	5/8	7
4	9	15/16	6 3/16	5 5/16	4.50	1 5/16	1 5/16	3	▲	4 5/8	4 1/32	8	5/8	7 1/2
5	10	15/16	7 5/16	6 7/16	5.56	1 7/16	1 7/16	3 1/2	▲	5 11/16	5 1/32	8	3/4	8 1/2
6	11	1	8 1/2	7 9/16	6.63	1 9/16	1 9/16	3 1/2	▲	6 3/4	6 1/16	8	3/4	9 1/2
8	13 1/2	1 1/8	10 5/8	9 11/16	8.63	1 3/4	1 3/4	4	▲	8 13/16	8	8	3/4	11 3/4
10	16	1 3/16	12 3/4	12	10.75	1 15/16	1 15/16	4	▲	10 15/16	10	12	7/8	14 1/4
12	19	1 1/4	15	14 3/8	12.75	2 3/16	2 3/16	4 1/2	▲	12 15/16	12	12	7/8	17
14	21	1 3/8	16 1/4	15 3/4	14.00	2 1/4	3 1/8	5	▲	14 1/4	13 1/4	12	1	18 3/4
16	23 1/2	1 7/16	18 1/2	18	16.00	2 1/2	3 7/16	5	▲	16 1/4	15 1/4	16	1	21 1/4
18	25	1 9/16	21	19 7/8	18.00	2 11/16	3 13/16	5 1/2	▲	18 1/4	17 1/4	16	1 1/8	22 3/4
20	27 1/2	1 11/16	23	22	20.00	2 7/8	4 1/16	5 11/16	▲	20 1/4	19 1/4	20	1 1/8	25
24	32	1 7/8	27 1/4	26 1/8	24.00	3 1/4	4 3/8	6	▲	24 1/4	23 1/4	20	1 1/4	29 1/2

All flanges are furnished faced, drilled, and spot-faced or back-faced.

††Minimum thicknesses of flanges include the 1/16-inch raised face.

‡‡See page 235 for dimensions of other standard facings.

*Dimension H is equal to the OD of pipe.

**Dimension A corresponds with the ID of standard pipe, also with schedule 40, sizes 12 inches and smaller, referred to in ASA B16e.

▲**Slip-on flanges** are bored to slip over pipe.

Bolt holes shall be drilled 1/8 inch larger in diameter than the normal size of the bolt.

Slip-on flanges are not chamfered or beveled at the flange face or end of hub unless so specified by customer.

Walworth Lap flanges will be furnished with the bore given above, unless other dimension is specified by customer.

Welding neck flanges are carried in stock with bores as listed. Sizes 12 inches and smaller can be furnished bored for extra heavy pipe and all sizes can be furnished bored for any thickness of pipe required.

Fig. 9.6. 150-Pound American Standard forged-steel flanges. (Data from ASA B 16e.) (Courtesy Walworth Co.)

WELDING

In modern industry, welding is a common method of assembling piping systems.

There was a time when designers believed that the use of threaded fittings on sizes 2 inches and smaller was cheaper than the use of welding fittings, but the improvement of welding techniques and the development of the socket-type fitting has led to the widespread use of welding fittings on pipe as small as ½ inch. In modern gasoline plant-construction practice, malleable-iron screwed fittings are acceptable only for domestic water supply or as fittings for instrument air. Nothing less than forged steel is recommended for all screwed couplings, and that only if the piping is subject to frequent removal. Forged-steel socket-welding fittings are recommended for all ½-inch and ¾-inch process and high-pressure steam services. The socket weld or butt weld may be used on 1½-inch pipe, and in the 2-inch sizes butt-welding fittings are commonly used.

It is not usually the draftsman's task to deal with problems of design, but as his responsibilities increase with his experience, there will be occasions when it will become his duty, on routine jobs to determine if welding should be used, and if so, how to combine fittings, pipe, valves, and special parts to accomplish the desired results in the space allowed. Consideration must be given to several points of design, such as the avoiding of sharp bends in order to prevent air pockets or turbulence in the line, adequate clearances for welding, and the possibility of later removal for repair or replacements.

The design and location of adequate supports and the orderly arrangement of the piping from the standpoint of appearance and convenience is also the responsibility of the draftsman.

In plants that have fabrication shops, the draftsman must also determine what portion of the system should be prefabricated and provide detail drawings for the prefabricated parts.

PREFABRICATED PARTS

Figure 9.7 is a reproduction of a shop drawing of a 90° bend. Note that the list of material includes

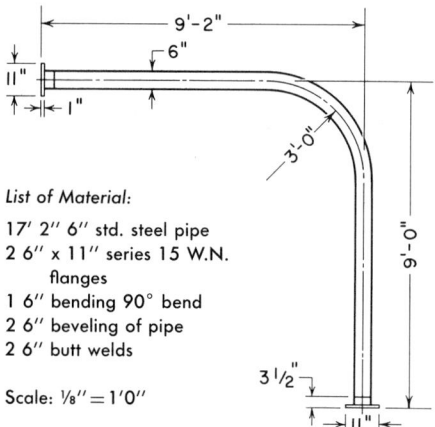

Fig. 9.7. Shop drawing for a 90° bend.

a list of the beveling and welding processes in addition to the listing of the actual pieces with descriptions of each piece. If screwed fittings are used, the actual cut length of each pipe should be listed along with the thread specifications.

When designing prefabricated parts the draftsman should consider whether he is to utilize combinations of standard welding fittings or specify pipe bends that can be formed in the shop.

WELDING SYMBOLS

The American Welding Society recommends the use of symbols for the representation of welds. These recommendations can be found in the Appendix of this text. For detailed instructions as to the use of welding symbols the student should refer to the publications of the American Welding Society, or to the *American Standard Graphical Symbols for Welding and Instructions for Their Use*, ASA Z32.2.1–1949 Reaffirmed 1953.

Problems

Problem 1. Figure 9.8 is a freehand single-line drawing of an open-float steam-trap assembly. The pipe sizes and the necessary dimensions for the trap and sediment separator are shown on the drawing. Make a double-line drawing in pencil on an 11 x 17 of tracing paper with attention to the following points:

1. Assembly must tie in to the existing system at the points of attachment as indicated on the drawing.

DETAIL DRAWING

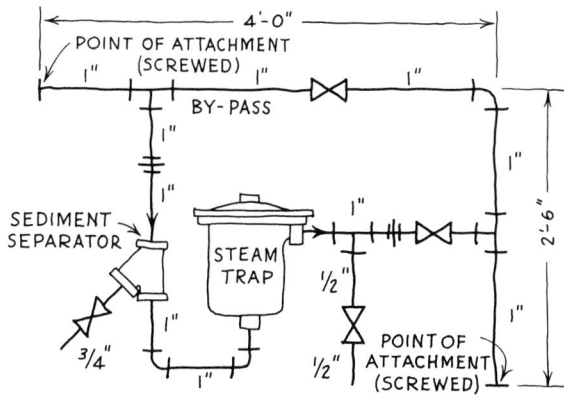

200# STEAM TRAP - HEIGHT 12½" - DIAMETER 9⅞"
SEDIMENT SEPARATOR - END-TO-END 5"

Fig. 9.8.

2. Arrange for location on the same sheet with the drawing a material bill which will provide for identification marks, description, and number required for all parts.
3. Label each pipe with an encircled upper-case letter for the purpose of identification on the material bill.
4. Show in the description in the material bill the cut length of each pipe, taking into consideration the thread engagement and allowances for fittings and valves.
5. Identical parts may be identified by like marks and the number required for all like parts specified in the material bill.
6. Use simplified double-line symbols.
7. Draw all center lines and dimension drawing by giving center-to-center dimensions of all parts.

Problem 2. Figure 9.9 is a freehand sketch of a control-valve assembly. A minimum of information has been given. Make an accurate and completely dimensioned double-line shop drawing with a material bill showing all information needed for the fabrication of one assembly.

Select a scale and sheet size to fit the needs of the problem.

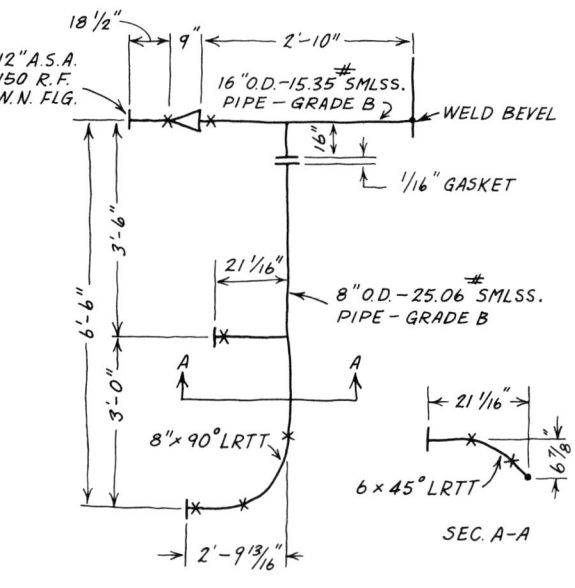

Fig. 9.9.

APPENDIX

The language of piping

Like all other craftsmen, industrial piping maintenance men have a language typical of their craft. Certain terms or words when used in the craft may have a different meaning from the one when they are used in everyday conversation. It is important that the draftsman be able to understand and use the language of piping.

ALL IRON. Term denoting that all parts of a valve are made of iron.

ANCHOR. A type of pipe support used to hold piping rigid at a given point.

ANGLE VALVE. A variant of the globe-valve design, having pipe openings at right angles to each other.

AUTOMATIC STOP-CHECK VALVE. Combination check and shutoff valve designed primarily for use on multiple boiler installations.

BAND. The raised collar put on the ends of certain screwed fittings and valves for reinforcement of ends.

BLIND FLANGE. Solid platelike fitting used to seal the end of a flanged end pipe line.

BLOW-OFF SYSTEM. Piping hookup used for blowing scale, sediment, etc., from boilers, tanks, or receivers.

BLOW-OFF VALVE. A valve designed specially for blow-off service and used in blow-off lines.

BRASS TO IRON. Designates a brass disc and iron seat, or vice versa, in a valve.

BRASS TRIM OR BRASS MOUNTED. Indicates that certain inside parts of a valve, such as stem, disc, seat rings, etc., are made of brass.

BUSHING. A threaded and tapped fitting which is used to reduce size of end opening of a valve or fitting.

BY-PASS. An auxiliary loop in a pipeline, usually for diverting flow around a valve.

CHECK VALVE. Valve designed to close automatically with reversal of flow in pipe line.

COCK. Original form of valve. Has tapered plug with hole which is rotated to provide passageway for fluids.

COMPOSITION DISC. Nonmetallic disc used in certain types of valves.

CONDENSATE. Liquid resulting from condensation of vapor or gas in a line.

CONVENTIONAL DISC. Most commonly used design of disc for glove valves.

CORROSION. Effect of deterioration of piping materials due to chemical action or electrolysis.

COUPLING. Female end pipe fitting used for joining two lengths of pipe.

CROSS. A cross fitting having four openings.

DISC. The part of a valve that actually closes off flow.

DRAINAGE FITTING. Type of fitting used mainly for plumbing drainage lines.

ELBOW. Fitting used for making a turn in direction of pipe line. Also known as "ell."

END CONNECTION. Refers to the type of connection by which piping elements are joined together.

EXPANSION JOINT. Pressure-tight device which permits expansion or contraction of pipe lines.

EXTRA STRONG: Denotes pipe sizes corresponding to Schedule 80.

FACING. Finish of the contact surface of flanged end piping materials.

FEMALE THREAD. Internal thread in pipe fittings, valves, etc., for making screwed connections.

FITTING. A piping element, other than valve or pipe, used in joining pipelines. Also the work of installing pipe as done by a pipe fitter.

FLANGE. Rim on the end of a pipe, valve, or fitting for bolting to another piping element.

FLANGED END. Refers to a valve or fitting having flanges for joining to other piping elements.

FOREIGN MATTER. General term for scale, rust, dirt, etc., in pipelines.

GATE VALVE. One of the basic valve types. Its name comes from its gatelike disc which regulates flow.

GLOBE VALVE. A basic valve type which gets its name from the globular shape of its body.

GROUND JOINT. Denoting a connection in which two machined metallic surfaces are joined face to face.

HANGER. A device for supporting a pipeline. It is made in various designs.

HEADER. A length of pipe or cast vessel to which two or more pipelines are joined to carry fluid from a common source to various points of use.

HICKEY. A length of pipe or extension handle used on a wrench to get greater leverage.

HUB END. Caulked or leaded type of end connection used on valves, fittings, and pipe mainly for water supply and sewerage lines.

INCREASER. Fitting with larger opening at one end, used to increase size of pipe opening. Common to drainage fittings.

JOINT. The point of connection between two piping elements whether screwed, bolted, welded, etc.

LAPPING-IN. Rubbing and polishing a surface such as disc face to obtain a smooth bearing with body seat rings.

LONG-SWEEP FITTING. Fitting with a long radius turn.

LUBRICANT. Specially prepared compound used when making up screwed joints to reduce friction and tearing of threads.

MALE THREAD. External thread on pipe fittings, valves, etc., for making screwed connection.

MALLEABLE FITTING. Pipe fitting made of malleable iron.

NEEDLE-POINT VALVE. Globe-type valve with needle-point disc for extremely fine regulation of flow.

NIPPLE. Short length of pipe (up to 12 inches long) threaded on both ends, for joining piping elements.

NONRISING STEM. Type of valve stem which merely turns and does not rise when valve is operated.

PACKING. Material used in stuffing box of a valve to maintain a leakproof seal around the stem.

PIPE DOPE. See Lubricant.

PIPE SCALE. A hard, scalelike material frequently found in new pipe. It is caused by heating operations in making pipe.

PIPE STRAP. Device for holding light-weight pipe to wall or ceiling.

PIPE SUPPORT. A device for supporting pipelines.

PIPING. General term for materials used in pipe lines. A complete piping system. Also the act of making up a pipeline.

PLUG. Screwed fitting for shutting off a tapped opening.

PLUG-TYPE DISC VALVE. Globe or angle valve with tapered plug disc and cone-shaped seat having wide bearing seating surface.

POP VALVE. Spring-loaded safety valve that opens automatically when pressure exceeds limits for which valve is set. Used as a safety device on boilers and other equipment to prevent damage from excessive pressure. Pop safety valves are generally used on steam, air, or other gases.

PORT OPENING. The pipe opening of a valve.

PRESSURE REGULATOR. A valve used to automatically reduce and maintain pressure below that of the source for certain processing or heating devices.

PRESSURE-TEMPERATURE RATING. Refers to "pressure-temperature rating as applied to piping."

RAILING FITTING. Type of fitting used for making up hand or guard railings.

RAILROAD UNION. Type of union. See union.

REDUCER. Fitting with smaller opening at one end for reducing size of opening.

REGULAR. Denotes a piping item regularly catalogued by manufacturer.

RELIEF VALVE. Safety valve similar to a pop safety valve in operation.

RETURN BEND. A "U" pattern fitting for reversing direction of pipe run. Used mostly in making up pipe coils.

RISING STEM. Type of valve stem that turns and rises when valve is open.

SCREWED END. Type of end on valve, fitting, or pipe that is joined to other elements by screwed connection.

SCREWED FLANGE. Flange that is attached to pipe by a screwed connection.

SIDE OUTLET. An ell or tee fitting with a side outlet.

SLIP-ON FLANGE. Flange which slips onto pipe and is welded in place.

SOCKET-WELDING FITTING OR VALVE. Socket-end-type valve or fitting that slips over end of pipe and is made pressure tight by welding.

SOFT JAWS. Copper or lead covers placed over vise jaws to prevent damage to materials held in vise.

SOLDER JOINT. Type of end connection made by soldering. Generally used with copper tubing.

SPECIAL. Denotes an item that differs from manufacturer's standard design.

STANDARD. Denotes pipe sizes corresponding to Schedule 40.

STREET ELL. Elbow fitting with male and female end.

TAP. Tool for forming internal or female thread.

TEE. A three-way fitting shaped like the letter "T."

THREADER. Tool for cutting male thread on end of pipe.

THROTTLING. Regulation of flow through a valve.

TRAP. Device for automatically draining condensate from lines.

TRIM. Term used in referring to bonnet, steam, disc, and seat parts of a valve.

TUBING. Light-weight pipe, usually copper, brass, steel, or plastic.

UNION. Fitting used to join lengths of pipe to permit easy opening of a line.

UNION FITTING. Fitting with union at one or more ends.

VALVE BODY. Main part of valve. Contains passageway for fluid and seating surfaces for discs.

WELDING END. Type of end on valve, fitting, or pipe that is joined to other piping elements by welding.

WELDING-NECK FLANGE. Flange with integral, extended neck for welding to pipe.

WHEEL. Wheel handle for operating a valve.

WIRE DRAWING. Term for the erosive effect upon face of valve seats usually due to cutting action of foreign particles in high-velocity fluids occurring especially when flow is closely throttled.

APPENDIX

63

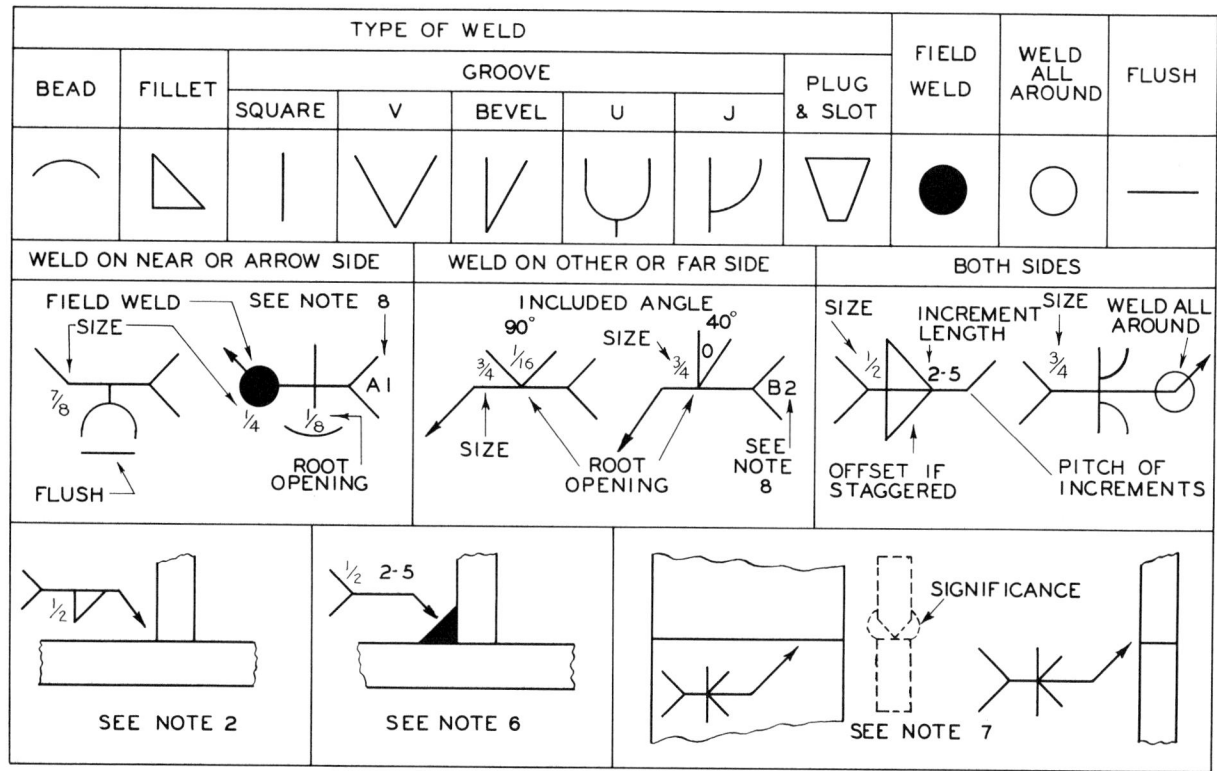

Arc and gas welding symbols recommended by the American Welding Society.

1. The terms "arrow side" and "other side" refer to the location of the nearest member parallel to the plane of the drawing.
2. If the arrow points to the end view or sectional view of a weld which is not drawn, the side closest to the arrow is the "near side."
3. Near and far welds are the same size unless otherwise shown.
4. Abrupt changes in direction of joint govern the limit of the application of the symbol unless otherwise dimensioned. Exception: When all around symbol is indicated.
5. All welds are continuous and of users standard proportions.
6. When welds are drawn in as in an end or sectional view, they are not indicated by symbol.
7. When only one member is grooved the arrow points to that member.
8. Specification reference appears in the tail of the arrow.

Index

American Standard, cast-iron flanges and flanged fittings, 12
    forged-steel flanges, 57
    screwed cast-iron fittings, 13, 35
    steel butt-welding fittings, 13
    taper pipe threads, 54, 55, 56
    wall thickness schedule, 35
American Standards Association, abbreviations, 42, 43
    for pipe threads, 53, 55
    allowable S values, 34
    recommendations for single line symbols, 4, 5, 6, 7
American Welding Society, recommendations for arc and gas welding symbols, 63
Anchors, 45
Appearance of assembly drawing, 48
ASME power boiler, allowable S values, 34, 35
Assembly, problems of, 53
Axis, axonometric, 19
    isometric, 19
Axonometric drawing, 18, 19

Bends, standard types, 52
Brackets, 44
By-pass, 21, 22

Center lines, use of, 10, 14
Clamp, I-beam, 43
Clearance, 48
Controls, problems, 36, 37
    special, 28, 29, 30

Detail drawing, 51
    problems, 58, 59
    scale of, 51
Developed views, 17, 18
    problems, 18
Diagram, flow, 20, 21
Diagram drawing, 20, 21
    problems, 21, 22
Dimensions of, butt-welding fittings, 13
    centrifugal pump, 32
    double-line symbols, 3, 8, 9, 10, 11
    flanged fittings, 12
    flanged valves, 11
    flanges, 57
    general arrangement drawings, 48
    motor, 32

Dimensions of, pipe, 35
    pipe threads, 54, 55, 56
    screwed fittings, 9, 13
    screwed valves, 10, 11
    single-line symbols, 3
    socket welding fittings, 36
Display drawings, 3
Drawing materials, 14, 15

End connections, flanged-end valve, 26, 27
    hub-end valve, 26, 27
    solder-end valve, 26, 27
    threaded-end valve, 26, 27
    welding-end valve, 26, 27
Equipment, use of, 11
    required for drawing, 14

Fabrication, pipe, 45, 51, 52, 58
    shop, 51
Fittings, cast-iron, flanged, 12, 36
    cast-iron, threaded, 13, 35
    forged-steel socket-welding, 36
    pipe, 12, 13, 35, 36
    specification of, 41
    steel butt-welding, 13
Flange dimensions, 57
Flange facings, 36, 38
Flanges, blind, 37
    lap-joint, 37
    reducing, 37
    service recommendations, 37
    slip-on, 37
    threaded, 37, 56
    welding-neck, 37
Flange types, 37, 55, 57
Flow, direction of, 10, 11
Flow diagrams, 20, 21
    problem, 22

Gage, liquid level, 28
Gasket, 10, 56
General arrangement drawing, 47, 48, 49
    problem, 49, 50
Governor, boiler fuel, 29

Hangers, pipe, 42, 43, 44

65

INDEX

Illustrations, catalog, 3
Inserts, concrete, 42, 44
Insulation, 45, 46
Isometric drawing, 18

Joints, threaded, 53, 55, 56

Layout, instructions for sheet, 14, 15
Lettering instructions, 14

Materials, pipe, 33
Materials, selection of, 41
Materials used in valves, 27, 30, 41
    brass, 27
    bronze, 27
    iron, 27
    steel, 27, 30
Motor dimensions, 32

Nominal pipe size, 33, 34, 35

Oblique drawing, 18
Orthographic projection, 17
    problem, 18

Parts list, 41, 42
Pictorial representation, 18
Pipe, selection of material for, 33
Pipe connections, double-line symbols for, 8, 12
    screwed, 8, 9
    welded, 13, 63
    flanged, 12, 36
    screwed, 13, 35
    single-line symbols for screwed, 4, 5, 6, 7
    welded, 4, 5, 6, 7
    welding, 13, 36
Pipe sizing, 33, 34
Piping draftsman, duties of, 1
Piping materials, characteristics of, 33
Piping systems, methods of representation, 17, 18
Position representation, 10, 14
Prefabricated parts, 58
Pump, dimensions of, 32

Regulator, back pressure or vacuum, 28
Rings, pipe, 43

Saddle, 44, 45
Sales drawing, 3
Scale recommendations, 3
Schedule numbers, 33, 34
Schematic drawing, 20, 21, 22
    problem, 22
Sediment separator, 28
Sheet layout, 14
Shop drawing, 51
Specification of parts, 41, 42, 43, 44, 45
    problem, 45
Specification of valves, 41
Steam trap, 28
Stem variations, 25, 26
Supplies needed for drawing, 14
Support for piping, 48
Supports, 44, 45
S values, allowable, 34

Symbols, accuracy of, 9
    double-line piping, 3, 9, 10
    conventional problem, 16
    procedure for drawing, 11, 14
    simplified double-line piping, 8
    problem, 16
    single-line piping, 3, 4, 5, 6, 7
    problem, 15, 16

Templates, 3
Thread engagement, 54, 55, 56
Thread representation, 53, 56
Threads, ASA recommendations for abbreviations, 53, 55
    ASA straight, 53, 55
    ASA taper, 53, 54, 55
    pipe, 53, 54, 55, 56
Thread specification, 53
Thread types, 53, 55
Title block, 15
Top works, reverse, 29
    types, 26
Tubing, 33

Valve, angle, 10, 11, 24
    dimensions of, 10, 11
    automatic stop-check, 25, 29
    check, 24, 25
    cock, 24, 25
    diaphragm control, 29, 30
    reverse top works, 29, 30
    weight loaded, 29, 30
    foot, 28
    free flow, 23
    gate, 24, 27, 28, 29
    double disc, 24
    wedge, 24
    globe, 23, 24, 27
    dimensions, 11
    disc, 23
    hose gate, 28
    lift check, 25, 27
    motor regulated, 29
    needle point, 28
    packless diaphragm, 29
    plug, 24, 25
    quick opening, 29
    relief, 24, 25, 28
    safety, 24, 25, 28
    swing check, 25
    throttling, 23
Valve designs, 23, 24, 25
Valve ends, flanged, 26, 27
    hub, 26, 27
    solder, 26, 27
    welding, 26, 27
Valve materials, brass, 27
    bronze, 27
    iron, 27
    steel, 27, 30

Welding, 58
Welding symbols, 63

Y valve, 29